Seven Theories of Human Nature

Seven Theories of Human Nature

Second Edition

LESLIE STEVENSON

New York Oxford
OXFORD UNIVERSITY PRESS

Oxford University Press

Oxford New York Toronto
Delhi Bombay Calcutta Madras Karachi
Petaling Jaya Singapore Hong Kong Tokyo
Nairobi Dar es Salaam Cape Town
Melbourne Auckland

and associated companies in
Beirut Berlin Ibadan Nicosia

Published by Oxford University Press, Inc.,
200 Madison Avenue, New York, New York 10016

Library of Congress Cataloging-in-Publication Data

Stevenson, Leslie Forster.
Seven theories of human nature.

Includes bibliographies and index.
1. Man. I. Title.
BD450.S766 1987 128 87-17170
ISBN 0-19-505291-9
ISBN 0-19-505214-5

6 8 9 7

Printed in the United States of America

TO MY PARENTS

Preface

I am gratified that this book has been so widely found useful at the introductory level for which it was designed. Given this success of the basic plan, I found no reason to change it for a new edition, so the text of the central seven chapters remains unaltered except for minor corrections and emendations. No doubt, almost every reader has a favourite suggestion for an eighth theory which should be added, but I found no overwhelmingly obvious candidate, and none that I wanted to drop, so the basic team remains. If it were to be augmented, I could see more reason to add another seven than to stop at one.

In the two opening chapters, I have taken the opportunity to amend some naiveties and infelicities, and in particular I have tried to be a little less superficial about the philosophy of science at the end of Chapter 2. There is, of course, vastly more to be said about each theory, but an introductory textbook is not the place to say it. To lead the reader towards deeper thought, I have taken some care to extend and bring up to date the recommendations for further reading at the end of each chapter. Bibliographies and reading lists are often too long and undiscriminating, it seems to me: I have mentioned only what I believe to be the best of the relevant work in each field, and I give some indication of the scope and level of each work.

The last chapter has been completely replaced. In the original edition I rather betrayed the applied, interdisciplinary character of the book by ending with a brief indication of some of the main problems recognized by academic philosophy (with a mere nod towards psychology and sociology). Somehow–no doubt because of my own employment–I seemed to be assuming that my readers would proceed to study philosophy in a more specialized way. While I still hope that some will do so, I now recognize that they will very likely be a minority. A greater number may want to ponder the mysteries of human nature at a less abstract level. So in the new final chapter I introduce a wider range of further questions and suggest a personal selection of readings.

Among these urgent questions is that of gender differences, which has been brought to our attention in recent years by the feminist movement. I have made no systematic attempt to 'desex' the language of this book, from the opening sentence 'What is man?' onwards–I hope readers will believe me when I say that I intend the masculine words to cover the whole human species. The issue seems to me to require a deeper response than that, yet it did not seem appropriate to extend the book by entering the debate, even if I were qualified to do so. What I have done is to indicate in the new Chapter 10 how this is one of several vitally important issues arising from this introductory discussion of human nature.

St. Andrews L. S.
May 1987

Preface to First Edition

This is an introductory book, intended simply as a rapid tour of a fascinating intellectual landscape. If it whets the reader's appetite for more detailed exploration, and helps him to start doing it for himself, then I shall have fulfilled my purpose. I assume no previous knowledge of the topics covered.

Librarians will find it hard to classify this book. Though written by a philosopher, it treats some writers and subjects not counted as philosophical in the academic sense. And though it considers some psychological theories, it could hardly count as a general introduction to psychology. It even strays into questions of biology, sociology, politics, and theology, thus overstepping the conventional faculty boundaries between arts, sciences, social science, and divinity. To use the word which is presently fashionable, it is 'interdisciplinary.' Perhaps it is best described as an extended exercise in what I have called 'applied philosophy' (in *Metaphilosophy* I: 3, July 1970, 258–67), that is, the application of conceptual analysis to questions of belief and ideology which affect what we think we ought to do, individually and socially. Inevitably, questions of pure philosophy are raised and not answered; I hope that some readers will be led into pursuing them further.

My thanks are due to my colleagues Keith Ward, Bob Grieve, and Roger Squires for their critical comments on parts of the manuscript, to my father Patric Stevenson for suggestions about style, to my students at the University of St. Andrews for their testing of my ideas and exposition, to Ena Robertson and Irene Freeman for efficient typing, and to my wife Pat for everything.

St. Andrews L. S.
October 1973

Contents

I

Introduction

1

Rival Theories

What is man? This is surely one of the most important questions of all. For so much else depends on our view of human nature. The meaning and purpose of human life, what we ought to do, and what we can hope to achieve—all these are fundamentally affected by whatever we think is the 'real' or 'true' nature of man.

The use of the masculine word 'man' here is very convenient for brevity of question and statement, and, as the quotations made in the next paragraph show, it has been very common practice. But straight away many of us will want to protest that what is involved is more than mere linguistic convenience, that some distinctive features and problems of women's nature have all too often been overlooked by the common assumption that the concept *man* can represent the whole human species. This book does not attempt any systematic discussion of feminist issues: it presents some rival theories of general *human* nature. Some readers may wish to pursue the implications for gender differences, and for this purpose some further reading is recommended at the end of Chapter 10 (note 5).

Even within the most masculine-oriented views of human nature, there are disagreements aplenty, and more than enough for this book to consider. 'What is man that Thou art mindful of him . . . Thou hast made him a little lower than the angels, and hast crowned him with glory and honour,' said the author of Psalm 8 in the Old Testament. The Bible sees man as created by a transcendent God who has a definite purpose for our life. 'The real nature of man is the totality of social relations,' said Marx (in his theses on Feuerbach in 1845). Marx denied the existence of God and held that each individual is a product of the human society he lives in. 'Man is condemned to be free,' said Sartre, writing in German-occupied France in the early 1940s. Sartre was as much an atheist as Marx, but (in that period of his thought, at least) he differed from Marx in holding that we are not determined by our society or by anything else. He held that every human individual is completely free to decide for himself what he wants to be and do.

Different views about human nature lead naturally to different conclusions about what we ought to do and how we can do it. If God made us, then it is His purpose that defines what we ought to be, and we must look to Him for help. If we are made by our society, and if we find that our life is somehow unsatisfactory, then there can be no real cure until society is transformed. If we are fundamentally free and can never escape the necessity for individual choice, then the only realistic attitude is to accept our situation and make our choices with full awareness of what we are doing.

Rival beliefs about human nature are typically embodied in various individual ways of life, and in different political and economic systems. Marxist theory (in one or another version) so dominates public life in most communist-ruled countries that any questioning of it can have serious consequences for the individual. In the so-called 'free' or 'democratic' nations we can easily forget that a few centuries ago Christian belief occupied a similarly dominant position: heretics and unbelievers were discriminated against, persecuted or burned. Even now, in some countries and in some areas, there is a socially established 'Christian' consensus which one can oppose only at some risk. In the Republic of Ireland, for example, Roman Catholic doctrine is constitutionally accepted as limiting policy on social matters such as abortion, contraception, and divorce. In the United States, there is an informal Christian ethos which affects the sayings (if not the actions) of politicians, despite the official separation of Church and State.

There is thus a tendency for the people and leaders of the superpower nations (U.S. and U.S.S.R.) to see themselves as in a competition that is not merely one of national rivalries, but of ideologies, each of which sees the other as based on a false and pernicious theory of human nature. This book will raise some sceptical and critical questions about both sides of this perilous confrontation.

An 'existentialist' philosophy like Sartre's may at first seem less likely to guide social practice; but one way of justifying modern 'liberal' democracy, with its separation of Church and State and its acknowledgment (as in the American Declaration of Independence) of the right of each individual freely to pursue his own conception of happiness, is by the philosophical view that there *are no* objective values for human living, only subjective individual choices. This assumption would seem to be incompatible with both Christianity and Marxism, but it is highly influential in modern Western society, far beyond its particular manifestation in French existentialist philosophy of the mid-century. It should be noted, however, that someone who believes there *are* objective standards may still support a liberal social system if he thinks it wrong to *enforce* them.

Let us look a bit more closely at Christianity and Marxism as two rival theories of human nature. Although they are radically different in content, there are some remarkable similarities in structure, in the way the parts of each doctrine fit together and give rise to ways of life. Firstly, they each make claims about the nature of the universe as a whole. Christianity is of course committed to belief in God, a personal being who is omnipotent, omniscient, and perfectly good, who created and controls everything that exists. Marx denied all this, and condemned religion as 'the opium of the people' which distracts them from their real social problems. He held that the universe exists without anybody behind or beyond it, and is fundamentally material in nature, with everything determined by the scientific laws of matter.

As part of their conception of the universe, both Christianity and Marxism have beliefs about the nature of history. For the Christian, the meaning of history is given by its relation to the eternal. God uses the events of history to work out His purposes, revealing himself above all in the life and death of Jesus. Marx claimed to find a pattern of progress in human history which is entirely internal to it. He thought that there is an inevitable development from one economic stage to another, so that just as feudalism had given way to capitalism, capitalism would give way to communism. Thus both views see history as moving in a certain direction, though they differ about the nature of the moving force and the direction.

Secondly, following from the conflicting claims about the universe, there are different descriptions of the essential nature of the individual human being. According to Christianity, he is made in the image of God, and his fate depends on his relationship to God. For each man is free to accept or reject God's purpose, and will be judged according to how he exercises this freedom. This judgement goes beyond anything in this life, for somehow each individual person survives the physical death that we know. Marxism denies any such survival of death and any such judgement. It must also deny the importance of that individual moral freedom which is crucial to Christianity, for according to Marx our moral ideas and attitudes are determined by the kind of society we live in.

Thirdly, there are different diagnoses of what is basically wrong with mankind. Christianity says that the world is not in accordance with God's purposes, that man's relationship to God is disrupted. He misuses his freedom, he rejects God, and is thus infected with sin. Marx replaces the notion of sin by that of 'alienation,' which conveys a similar idea of some ideal standard which actual human life does not meet. But Marx's idea is of alienation from oneself, from one's own true nature, since men have potential that the conditions of capitalist society do not allow them to develop.

The prescription for a problem depends on the diagnosis of the basic cause. So, fourthly, Christianity and Marxism offer completely different answers to the ills of human life. The Christian believes that only the power of God Himself can save us from our state of sin. The startling claim is that in the life and death of the particular historical person Jesus, God has acted to redeem the world and restore men's ruptured relationship with Himself. Each individual needs to accept this divine forgiveness, and can then begin to live a new regenerate life in the Christian church. Human society will not be truly redeemed until individuals are thus transformed. Marxism says the opposite—that there can be no real change in individual life until there is a radical change in society. The socioeconomic system of capitalism must be replaced by that of communism. This revolutionary change is inevitable, because of the laws of historical development; what the individual should do is to join the revolutionary party and help shorten the birth pangs of the new age.

Implicit in these rival prescriptions are somewhat differing visions of a future in which man is totally regenerated. The Christian vision is of man restored to the state that God intends for him, freely loving and obeying his Maker. The new life begins as soon as the individual accepts God's salvation and joins the Church, the community of the redeemed. But the process is only completed beyond this life, for both individual and community will still be imperfect and infected with the sin of the world. The Marxist vision is of a future in this world, of a perfect society in which men can become their real selves, no longer alienated by economic conditions, but freely active in cooperation with each other. Such is the goal of history, although it should not be expected immediately after the revolution, since a transitional stage will be needed before the higher phase of communist society can come into being.

We have here two systems of belief which are total in their scope. Both Christians and Marxists claim to have the essential truth about the whole of human life; they assert something about the nature of all men, at any time and in any place. And these world views claim not only assent but also action; if one really believes in either theory, one must accept that it has implications for one's way of life.

As a last point of comparison, note that for each belief-system there is a human organization which claims the allegiance of believers and asserts a certain authority on both doctrine and practice. For Christianity there is the Church, and for Marxism the Communist Party. Or to be more accurate, there are now many Christian churches and many Marxist parties, making competing claims to follow the true doctrine of their founder, defining various versions of the belief as orthodox, and following different practical policies. Such sect-formation is typical of both beliefs.

Many people have noted this similarity in structure between Christianity and Marxism, and some have suggested that the latter is as much a religion as the former. There is food for thought here for believers of both kinds, and for the uncommitted person too. Why should such very different accounts of the nature and destiny of man have such similar structures? Perhaps the differences can be reconciled to some extent, for there are those who claim to be Christian Marxists. But in the traditional interpretations of each belief, there are very basic disagreements about the existence of God and the nature of man.

But, as I have already suggested by quoting Sartre, there are many more views of man. The theories of the ancient Greeks, especially of their great philosophers Plato and Aristotle, still influence us today. More recently, Darwin's theory of evolution and Freud's psychoanalytic speculations have permanently changed our understanding of ourselves. Modern biology, psychology, and sociology offer a variety of allegedly scientific theorizing about human nature. Many distinguished scientists, including some to be mentioned in this book, have been ready to offer their own diagnosis of, and prescription for, the human condition, supposedly based on their own particular scientific expertise. Outside the Western tradition, there have been Chinese, Indian, African, and Islamic views of man, some of which are still very much alive. Islam in particular is undergoing a resurgence of popular strength, as the peoples of the Middle East express their rejection of many aspects of Western culture.

Some of these views are embodied in human societies and institutions and ways of life, as Christianity and Marxism are. If so, they are not just theories, but ways of life, subject to change and to growth and decay. A system of belief about the nature of man which is thus held by some group of people as giving rise to their way of life is standardly called an 'ideology.' Christianity and Marxism are certainly ideologies in this sense, and even value-subjectivism can, as we have seen, be a basis for political liberalism.

An ideology, then, is more than a theory, but is based on a theory of human nature which somehow suggests a course of action. What I want to do in this book is to examine certain influential theories which thus prescribe action as well as claim belief. Not all of them are ideologies, since not all have a corresponding group of people who hold the theory as giving rise to their way of life. But the theories I have selected to discuss all exhibit the main elements of that common structure we have seen in Christianity and Marxism: (1) a background theory of the nature of the universe; (2) a basic theory of the nature of man; (3) a diagnosis of what is wrong with man; and (4) a prescription for putting it right.

Only theories that combine such constituents offer us hope of solutions to the problems of mankind. For instance, the single assertion that all men are selfish is a diagnosis, albeit a brief one, but it offers no understanding of why we are selfish and no suggestion as to how we can overcome it. Similarly, the prescription that we should all love one another gives no explanation of why we find it difficult. The theory of evolution, although it has a lot to say about man and his place in the universe, does not in itself give any diagnosis or prescription.

The theories I am going to examine include those of Christianity, Marx, and Sartre. To them I add those of Plato's *Republic* (one of the most influential books of all time and one of the most readable works of Greek philosophy), of Freud (whose psychoanalytical theories have affected so much of the thought of this century), of B. F. Skinner (the American professor of psychology who claims to have the key to the problems of human behaviour), and of Konrad Lorenz (the Austrian Nobel Prize winner who has given new direction to the study of animal behaviour, and has led a recent fashion of explaining human nature by analogy with apes and other animals). In each case I cannot hope to trace the many antecedents of each view, although I shall try to sketch the essential background briefly. Nor can I survey the many varieties of Christian, Marxist, existentialist, and psychoanalytic theory. I shall simply try to introduce the key ideas of each, interpreting it through the four-part structure outlined above. In each case I shall take one readily available book as my basic text, and shall make references to it so that the reader can check my assertions and find out more for himself. I shall also suggest further reading relevant to each theory. Some readers will be disappointed that I do not discuss any Eastern views. To them I plead ignorance and shortness of time, and recommend some reading described below.

But as well as expounding the basic ideas, I want to suggest some of the main difficulties in them. So for each theory there will be some critical discussion which will, I hope, encourage the reader to think further for himself. In some chapters the critical questions will follow the exposition, in others they are intertwined with it. Before we begin our main task let us take another preliminary look at the cases of Christianity and Marxism, in order to see what may happen when we begin to criticize theories of human nature.

For Further Reading

Christianity and Marxism will be discussed in more detail later in this book, and further reading on each will be suggested. But for the comparison between them as belief-systems, the following can be recommended: *Philosophy and Myth in Karl Marx,* by Robert C. Tucker (Cambridge University Press, Cambridge, 1st edn. 1961, 2nd edn. 1973; see especially the introduction), and *Marxism and Christianity,* by Alasdair MacIntyre (Penguin Books, Harmondsworth, 1971; Schocken Books, New York, 1969).

For more on the notion of ideology, see *Ideology,* by John Plamenatz (Macmillan, London, 1971; Praeger, New York, 1970).

For introductions to Jewish, Chinese, Indian, and Islamic theories of human nature, see *The Concept of Man,* edited by S. Radhakrishnan and P. T. Raju (George Allen & Unwin, London, 2nd edn. 1966; Johnsen Publishing Co., Lincoln, Nebr., 1966). There are brief excerpts from some Indian and Chinese classics in Part I of

The Study of Human Nature, edited by Leslie Stevenson (Oxford University Press, New York, 1981). This anthology uses about half its length for readings representing the seven theories discussed in this book. *The Nature of Man,* edited by Erich Fromm and Ramon Xirau (Macmillan, New York, 1968) contains brief excerpts from a very wide range of thinkers, from many cultures and periods. The views of Aristotle, Hobbes, Adam Smith, Marx, Durkheim, Weber, and Schutz are discussed in Tom Campbell's *Seven Theories of Human Society* (Oxford University Press, New York, 1981) which follows the pattern of this book exactly (at twice the length).

2

The Criticism of Theories

The basic Christian assertion about the universe, that God exists, is of course faced with many sceptical objections. To take one of them, the evil and suffering in the world seems to count against the existence of God. For if He is omniscient He must know of the evil, and if He is omnipotent He must be able to remove it, so if He is perfectly benevolent why does He not do so? In particular, why does God not answer the prayers of believers for the relief of the manifold sufferings all over the world?

The basic Marxist assertion about the universe, that there is an inevitable progress in human history through stages of economic development, is just as open to scepticism. Is it really at all plausible that such progress is inevitable, does it not depend on many noneconomic factors which are not predetermined? In particular, communist revolutions have not occurred in the industrialized countries of Western Europe, so is this not direct evidence against Marx's theory?

Christian and Marxist claims about the nature of individual persons immediately raise large metaphysical problems. Is the individual really free and responsible for his actions? Or is everything about him determined by his heredity, upbringing, and environment? Does the individual person continue to exist after death or not? In the face of the universal and obvious fact of human mortality, the alleged evidence for survival is slim and highly controversial. But can the materialist view that men are made of nothing but matter really be true, in the light of our obvious mental powers to perceive and feel, to think and reason, to debate and decide?

Doubts also arise about the respective prescriptions for man's problem. The Christian claim that a particular historical figure is divine, and is the means of God's reconciliation with the world, defies all human rationality. The Marxist belief that communist revolution is the answer to the problems of humanity attaches almost as great a significance to a particular historical event. In neither case is the cosmic claim supported by the subsequent history of those individuals, institutions, or nations in which the regeneration is supposed to be taking effect. For the histories of the Christian Church down through the ages and of Russia since 1917 show a mixture of good and evil like all other human history. The practice of Christian or communist life has not eliminated muddle, disagreement, selfishness, persecution, tyranny, torture, and murder.

These common objections to each ideology are pretty well worn by now. What is interesting is that neither belief has disappeared in the face of them. Admittedly, Christianity has suffered a steady erosion of influence over the last few centuries; and perhaps only a small proportion of the population of the communist countries could count as convinced believers in Marxist theory. But both theories are still very much alive, for there are many Christian believers and many Marxist believers, each on both sides of the iron curtain. They have not vanished from the industrialized countries, in the way that witchcraft and astrology have vanished, except from the inside pages of the Sunday newspapers!

How and why is it that a significant number of non-lunatics continue to believe in Christianity or Marxism? Firstly, the believers usually find some way of explaining away the standard objections. The Christian says that God does not always remove evil, or answer our prayers, for what may seem bad to us may ultimately be for the best. The Marxist may say that revolution has not occurred in the West because the workers have been 'bought off' by the concession of higher standards of living, and have not realized that their true interest is in the overthrow of capitalism. Disputes about the great metaphysical questions of determinism or free will, materialism or immortality, seem able to go on forever without dislodging any side from its position. To the doubts about the respective prescriptions, the believers can reply that the full regeneration of man is still to come, and that the terrible things in the history of Christianity or communism are due to perfection being not yet achieved. By thus explaining away difficulties in his theory and appealing to the future for vindication, the believer can maintain his belief with some show of plausibility. The theorists of Church and State become well practised at such justification of the ways of God, or of the Party.

Secondly, the believer can take the offensive against criticism, by attacking the motivations of the critic. The Christian can say that those who persist in raising intellectual objections to Christianity are being blinded by sin, that it is their own pride and unwillingness to receive the grace of God that prevents them from seeing the light. The Marxist can similarly say that those who will not recognize the truth of Marx's analysis of history and society are being deluded by their 'false consciousness,' the ideas and attitudes which are due to their economic position: the capitalist mode of production naturally prevents those who benefit from it from consciously acknowledging the truth about their society. So in each case, a critic's motivations can be analysed in terms of the theory he is criticizing, and the believer may therefore think he can dismiss the criticism as based on illusion.

These are two of the main ways in which a belief can be maintained in the face of intellectual difficulties. If a theory of human nature is maintained by these two devices–(1) of not allowing any conceivable evidence to count against the theory, and (2) of disposing of criticism by analysing the motivations of the critic in terms of the theory itself–then I shall say that the theory is being held as a 'closed system.' It appears from the above that Christianity and Marxism can be held as closed systems–but this is not to say that all Christians or Marxists hold their belief in that way.

Why should people maintain a belief in the face of difficulties? Inertia, and unwillingness to admit that one is wrong, must play some part here. If one has been brought up in a certain belief and way of life, or if one has been converted to it and then followed it, it takes courage to abandon one's past. When a belief is an ideology, giving rise to the way of life of a social group, it will always be difficult for the members of the group to consider it objectively. There will be strong social pressures to continue to acknowledge the belief, and it will be natural for believers to maintain it as a closed system. People will feel that their belief, even if open to objections, contains some vital insight, some vision of essential truths. To abandon it may be to abandon what gives meaning, purpose, and hope to one's life.

Is it possible then to discuss various theories of human nature rationally and objectively, as I am setting out to do in this book? For when such theories are embodied in ways of life, belief in them seems to go beyond mere reasoning. Indeed, it can make itself apparently impregnable to criticism by the above devices of the closed system. The ultimate appeal may be to faith or authority, and there may be no answer to the questions 'Why should I believe this?' or 'Why should I accept this authority?' which will satisfy someone who is not already inclined to believe. The project of this book may therefore seem to be doomed from the start, if we jump to the conclusion that there can be no objective dicussion of rival ideologies.

However I believe that such despair would be premature. For one thing, not all the theories I am going to discuss are ideologies at all, and when they are not there is much less likelihood of their being believed and defended in this closed-minded way. But more importantly, even when a belief becomes an ideology and is perhaps held as a closed system by some believers, I think we can see that rational discussion is still possible for those who are prepared to try it. For we can always distinguish what someone says from his motivation for saying it. The motivation may be important in various ways, for instance if we wish to understand the personality of the speaker and the nature of his society. But if we are primarily concerned with the truth or falsity of what is said, and with whether there are any good reasons for believing it, then motivation is irrelevant. The reasons that the speaker may offer are not necessarily the best reasons. There is nothing to stop us discussing what he says purely on its own merits.

This is why the second feature of closed systems—the technique of meeting criticism by attacking the motivation of the critic—is fundamentally irrational. For if what is being discussed is whether the theory is true, or whether there are good reasons for believing it, then the objections that anyone produces against it must be replied to on their own merits, regardless of their possible motivations. Someone's motivation may be peculiar or objectionable in some way, and yet what he actually says may be true, and justifiable by good reasons. Even if motivation *is* to be considered, to analyse it in terms of the theory under discussion is to assume the truth of the theory, and therefore to beg the question. An objection to a theory cannot be defeated just by reasserting part of the theory.

The first feature of closed systems, the explaining away of all evidence against the theory, must also be looked at with some suspicion. We often feel that such 'explaining away' is not really very convincing, except to someone who is already disposed to believe in the theory. (Consider how Christians may answer the problem of evil, and Marxists the problem of why no revolutions have occurred in the West.) We must try to see when such explaining away is rationally justifiable, and when it is not. To do this, we must decide what *sort* of statement is being made, before we can discuss the relevance of alleged evidence for or against it.

Firstly, a statement may turn out to be a value judgement, saying what *ought* to be the case, rather than a statement of fact, about what *is* the case. For example, suppose someone says that homosexuality is unnatural. It might be objected to him that in almost every known human society there is a certain amount of homosexuality. Suppose he replies that this does not disprove what he says, since it involves only a minority in each society. Perhaps the objector will suggest that it is possible that a majority of society might indulge in homosexual as well as heterosexual activity (and that this seems to have been the case in ancient Greece). The reply may be 'I would *still* say it is unnatural.' Such a reply suggests that the speaker is not after all asserting anything about what people actually do, but is expressing an opinion about what they ought to do (or ought *not* to do!). This impression would be confirmed if we find that the speaker reacts with horror against anyone described as homosexual. If what is being asserted is thus really 'evaluative' and not factual, then evidence of what actually happens does not disprove it, for it is perfectly consistent to say that what does happen should not happen. But in order for the statement to be rightly allowed to be thus impervious to evidence, it must be recognized as a value judgement, as not even *attempting* to say what *is* the case. And if so, then it cannot be *supported* by evidence either, for what actually happens is not necessarily what should happen.

Statements about human nature are especially subject to this kind of ambiguity. Indeed, the words 'nature' and 'natural' should be regarded as danger signals, indicating possible confusion. If someone says 'Human beings are naturally X,' we should immediately ask him 'Do you mean that all or most human beings *are* actually X, or that we *should* all be X, or what?' Maybe what is meant is 'Whenever human beings are not X, they suffer consequences Y.' Here we have *both* a factual generalization and an implicit value-judgement about the undesirability of Y, and it will be appropriate to ask for evidence for the former, and for reasons for the latter. The objectivity of value-judgements is one of the central questions of philosophy, and I am not prejudging it here. I am just pointing to the need for these kinds of clarifying questions when discussing human nature.

There is a second, quite different, way in which a statement may correctly be held to be impregnable to contrary evidence, and that is if it is matter of definition. For instance, if someone says that all men are animals, it is not clear how any conceivable evidence could count against him. Suppose that the theory of evolution were not true, that it were found that we do not after all have a common ancestry with any other species.

Would we not still count as animals, albeit a special kind of animal, since we live, feed, breed, and die like all other animals? Suppose robots were made to walk and talk like men, but not to eat or reproduce like us. Clearly they would not be animals, but could they count as men either? It looks as if nothing could be *called* a man unless he could also be called an animal. If so, the statement that all men are animals does not really make any assertion about the facts about men, but only reveals part of what we mean by the word 'man.' It is true by definition, true in virtue of meaning alone. In philosophers' terminology, it is 'analytic' (it can be shown to be true by analysis of the meaning of its terms). If a statement is thus analytically true, it is quite correct to say it cannot be refuted by any conceivable evidence, but neither of course can it be proved by evidence, for it does not *attempt* to say anything about the state of the world.

The example of 'All men are animals' shows that a statement which appears to be saying something about the facts of human nature may really be only a concealed definition. Not all matters of definition are trivial, however. If a word is already used with a standard meaning in the language, it will be extremely misleading for anyone to use it with a different meaning, unless they give us explicit warning. Sometimes theories introduce new terms, or use old words in new ways, and it will then be very necessary for definitions to be given, and for it to be made clear that they are definitions, not claims about any sort of fact. And definitions may have consequences which are not immediately obvious, for instance if it is analytic that all men are animals and that all animals feed, then it is analytic that all men feed. Analytic statements, then, can have their uses, but only if they are clearly distinguished from 'synthetic statements' which make genuine assertions about the facts. There has been a debate among philosophers about whether this distinction is as clear as it seems at first, and even about whether there ultimately is any such distinction. But without entering into that difficult theoretical question here, I think we can see that if someone maintains that all men are X and dismisses without investigation any suggestion that some men might not be X, then we must ask him 'Is it part of your definition of a man that he must be X, or would you allow the conceivability of some man being discovered not to be X?' Only if he admits it to be a matter of definition can he be allowed to dismiss evidence without further investigation.

Value judgements and analytic statements, then, are not the sort of statements which can be proved or disproved merely by investigating the facts about the world. If a statement *can* be supported or shaken by such investigation–ultimately involving what someone can observe using his senses of sight, touch, sound, smell, and taste–it is called by philosophers an 'empirical' statement. By use of the above clarifying questions it should usually be possible to elucidate whether a statement is evaluative or analytic rather than empirical. The really difficult cases are when a statement does not seem to fall into any of these three categories. Consider again the Christian assertion of the existence of God and the Marxist assertion of an inevitable progress in history. It is pretty clear that these statements are trying to say something about what is the case, to assert some fundamental truth about the nature of the universe. Their proponents will hardly admit them to be value judgements or mere matters of definition. Yet it is not clear that these assertions are genuinely empirical either, for as we have seen above, although there would seem to be a very great deal of evidence which might well be thought to count against each one, their proponents often do not accept this as contrary evidence but find ways of explaining it away. Now if a believer in a theory seems ready to explain away *all possible* evidence against it (making additions to his theory if necessary), we begin to feel that he is winning too easily, that he is somehow breaking the rules of the game. How can a statement really assert something about the facts unless it is in principle open to testing by observation of some kind?

This is why some philosophers have felt attracted to what has been called 'the verification principle,' which stated that no nonanalytic statement can be meaningful unless it is verifiable by observation (value judgements were dismissed as not really statements at all, only expressions of emotion). If one accepts this principle, one will dismiss any so-called 'metaphysical' statement, which is neither analytic nor empirical, as not just false but meaningless. The questions of the existence of God, and of an inevitable progress in history, and many others (including ones more directly about human nature, such as the existence of an immortal soul) were indeed dismissed as meaningless by the 'logical positivists' (as the proponents of the verification principle were called). Yet many others have thought that this is too short a way with such big questions, and so one of the major philosophical debates in this century has been about whether any such verification principle should be accepted.

In so far as any conclusion has emerged from that debate, it is that although it is very important to distinguish statements that are either analytic or empirical from those that are neither, we cannot dismiss all the latter as meaningless. They are too mixed a bag, and many deserve individual attention. The stark choice between meaningful and meaningless is surely too crude a tool with which to explore the problems involved in statements about the existence of God, progress in history, or the immortality of human souls. Such claims are not nonsense in the way that 'The mome raths outgrabe' is, nor in the (different) way in which 'Green ideas sleep furiously' is; nor are they explicitly self-contradictory like 'Some leaves are both green and colourless.' However, the challenge remains that any statement which is neither a value judgement nor analytic, and which does not seem to be testable by observation either, is still deeply problematic in status. What reasons are there for accepting any such claims–how could any of them offer us *knowledge* rather than empty speculation?

Philosophers of science have tried to elucidate what it is that apparently enables *scientific* theories to give us reliable knowledge about the nature of the world. Certainly, science must depend on statements about what observably happens, for example, when some experiment is conducted, but scientific theories also make claims about what we *cannot* perceive, whether because it is too faraway in space, or in the distant past, or too small for any human senses to detect. How can these claims command rational assent—in what way are they any better off than 'metaphysical' statements?

Typically, it is because they can be tested indirectly—they have consequences (in conjunction with other assumptions) which *can* be observed. The distinguished philosopher of science Karl Popper put the emphasis on falsification rather than verification here, holding that the essence of scientific method is that theories are *hypotheses,* which can never be known for certain, but which are deliberately put to the test of observation and experiment, and revised or rejected if their predictions get falsified. Conclusive falsification, beyond all possible doubt, may not be achievable—as philosophers of science since Popper have had to acknowledge—but the main point is that observable evidence *can* rationally count against a scientific theory. So when I talk of 'falsification' later in this book, what I have in mind is not conclusive falsification, but some empirical observations providing, in the light of presently available considerations, evidence against a claim.

The examples already mentioned suggest that some controversial statements about human nature may not be held as scientifically testable hypotheses at all. This need not condemn them outright, but it is a very important feature to establish, for they cannot then enjoy any of the advantages of scientific status—that their defenders can point to the observable evidence and the connecting arguments, and challenge anyone how they can rationally reject the claims. Conceivably, there may be other sorts of reasons for accepting them, but we had better inquire carefully what they are, in each case.

I think this is about as much as we can usefully do by way of general preparatory methodology here, so let us now begin our critical examination of particular theories of human nature.

For Further Reading

My use of the notion of a 'closed system' is derived from that of Arthur Koestler, in *The Ghost in the Machine* (London, 1967; Pan paperback, 1970; Regnery Gateway paperback, Chicago, Ill.), p. 300. (This book contains many interesting but highly controversial assertions about human nature, on topics discussed in Chapters 8 and 9 of this book.)

For the classic statement in English of the verification principle, and of the meaninglessness of ethical and theological statements, see A. J. Ayer, *Language, Truth, and Logic* (first published in 1936, now in Penguins; Dover paperback, New York). For a more recent introductory discussion of the status of the verification principle, see W. P. Alston, *Philosophy of Language* (Prentice-Hall, Englewood Cliffs, 1964), Ch. 4.

The falsifiability criterion of a theory's being scientific is due to Karl Popper. See his book *The Logic of Scientific Discovery* (first published in 1934, now available in paperback from Hutchinson—London, 1959, revised edn. 1968; Harper & Row Torchbook paperback, New York), especially Chapters I–V. (Later parts of the book become highly technical.) For an easier introductory discussion, see Bryan Magee, *Popper* (Modern Masters Series, Fontana, London, 1973; Viking, New York, 1973), and A. F. Chalmers, *What Is This Thing Called Science?* (Milton Keynes, The Open University Press, 2nd edn. 1982).

II

Seven Theories

3

Plato: The Rule of the Wise

Let us start our examination of rival theories of man by considering that of Plato (427–347 B.C.) as an example of the fourfold pattern of claim about the universe, claim about human nature, diagnosis, and prescription. Although so old, it is still of contemporary relevance, for, whenever anyone asserts that the cure for our problems is that we should be ruled by those who really know what's best, then he is asserting the essence of Plato's theory.

A short sketch of Plato's background will help us to understand the origin of his ideas. He was born in the Greek city-state of Athens, which had for some time enjoyed economic prosperity through its trade, democratic government in the time of Pericles, and unprecedented advances in intellectual inquiry culminating in the great ethical philosopher Socrates. But Plato grew up in a period of war, which ended in defeat for Athens and a brief period of tyranny. When democracy was restored Socrates was condemned to death on a charge of impiety and corrupting the youth. Socrates' teaching was akin to that of the Sophists, who claimed to teach the art of rhetoric or persuasion, an art which was particularly useful in Athenian democracy. The Sophists also discussed moral and political theory, and among the opinions commonly expressed was a skepticism about whether any moral or political rules were more than arbitrary conventions, in view of the different practices in different societies (known to the Athenians through trade). Socrates' main concern was with how we can know the right way to live, and in this he much influenced Plato, who was deeply shocked at the execution of his teacher for his allegedly subversive questioning of conventional opinions. Disillusioned with contemporary politics and philosophy, Plato sought both knowledge of the truth about the universe, and the cure to the ills of society. The conclusions he reached are put into the mouth of Socrates in the many philosophical dialogues which Plato wrote, and were taught in the Academy he founded, which was in effect the world's first university.

Deservedly the most famous of Plato's dialogues is the *Republic,* in which he outlines his conception of the ideal human society. In the course of the book, he gives his view on many topics, including philosophy, morals, politics, education, and art. It is mainly this dialogue that I will consider here, and I will use the traditional numbering system (reproduced in nearly all editions and translations) for references to the text. I will first expound the main doctrines, and then criticize them in turn.

Theory of the Universe

Although Plato mentions God, or the gods, at various places, it is not clear how seriously he takes them, whether singular or plural. When he does talk of God in the singular, it is pretty clear that he does not mean anything like the personal God of Christianity, and even this impersonal notion of God does not play much of a role in the argument of the *Republic.*

What is really central to Plato's concept of the universe is his theory of Forms. This can be summarized under four aspects–logical, metaphysical (to do with what is ultimately real), epistemological (to do with what can be known), and moral. How is it that one word–for example, 'cat'–can truly apply to many different individual things? Plato's answer is that corresponding to each such general word there is one Form–in this case, the Form Cat–which is something different from all the individual cats (596). What makes these particular animals cats is their resemblance to, or 'participation in,' the Form Cat. This is the logical aspect of the theory–an answer to a question about the meaning of general words, the so-called 'problem of universals.'

The metaphysical aspect is that these Forms are held to be more real than material things, for they do not change or decay. The Forms are not in space or time, and they are not perceivable by any of the senses (485, 507, 526–7). Plato's vision was that beyond the world of changeable and destructible things there is another world of unchanging eternal Forms. The things we can see and touch are only very distantly related to these ultimate realities, as he suggested by his famous comparison of the human condition with that of prisoners chained facing the inner wall of a cave, so that all they can see are mere shadows of objects in the cave, knowing nothing of the world outside (515).

However Plato thought that by a process of education it is possible for human minds to attain knowledge of the Forms. The epistemological aspect of the theory is that only this intellectual acquaintance with the Forms can really count as knowledge, since only what fully exists can be fully known. Perception of impermanent material things is only belief or opinion, not knowledge (476–80).

The most convincing illustration of these three aspects of the theory of Forms comes from the Euclidean geometry with which Plato was familiar. Consider how it deals with lines, circles, and squares, although no physical object is perfectly straight, circular, or square but may always have some irregularity. Theorems concerning these ideal objects—straight lines without thickness, perfect circles, etc.—are proved with absolute certainty by logical arguments. So here we have indubitable knowledge of timeless objects which are the patterns that material objects imperfectly resemble.

It is the moral aspect of the theory of Forms that plays the most important role in Plato's theory of human nature and society. Consider moral words such as 'courage' and 'justice': as for all general terms, Plato will distinguish the many particular actions which are courageous or just, and the many different individual people who might be said to be courageous or just, from the Forms Courage and Justice. The general word is true of just those actions or people who in some way exemplify the corresponding Form. And rather as in the geometrical examples, no action or person is an absolutely perfect example of courage or justice, because of the truism that nobody is morally perfect. So the moral Forms set the objective moral standards by which human conduct and character should be judged. The word 'good' is the most general moral word, so the Form Good is pre-eminent among the Forms, and plays an almost God-like role, being the source of all reality, truth, and goodness (505–9). The absolute standards set for us by the Forms are not just for individuals, but for the whole of social and political life, they define the ideal form of human society (472–3). The theory of Forms, then, is Plato's answer to the intellectual and moral scepticism of his time. It is one of the most persuasive statements of the power of the human intellect to attain genuine knowledge about the universe and about the goals of human life and society. It can be seen as the culmination of Greek confidence in the intellect and Socratic concern with ethics.

Theory of Human Nature

Plato is one of the main sources for the 'dualist' view of man, according to which the soul or mind is a non-material entity which can exist apart from the body. He maintained that the human soul is indestructible, that it has existed eternally before birth and will exist eternally after death. These doctrines are stated in the *Republic* (608–11), but Plato's main arguments for them are given in other dialogues, especially the *Meno* and the *Phaedo*. The doctrine of the immateriality and immortality of the soul is not central in the *Republic,* but it goes naturally with Plato's contrast of the world of Forms with the world of perceivable things. For he held that it is the soul, not the body, which attains knowledge of the Forms, and which is the concern of ethics.

More central to the argument of the *Republic* is the doctrine of the three parts of the soul (435–41). Consider cases of mental conflict, such as when someone is very thirsty but does not drink the available water because he knows it is poisoned. Plato argues that there must be one element in the person's mind which is bidding him drink, and a second which forbids him; the first is called desire or Appetite (by which is intended all the physical desires, such as hunger, thirst, and sexual desire), and the second is called Reason. The existence of the third element in the mind is proved, Plato thinks, by other cases of mental conflict where a person feels angry or indignant with himself, for instance in the story he tells of the man who felt a fascinated desire to look at a pile of corpses and yet was disgusted with himself for wanting to. Plato maintains that what is in conflict with his Appetite here is not Reason but a third element which he variously calls indignation, anger, or Spirit. He thinks that children show Spirit long before they display reasoning; it is something like self-assertion or self-interest, and is usually on the side of Reason when inner conflict occurs. Reason, Spirit, and Appetite are present in every person, but according to which element is dominant we get three kinds of men, whose main desire is, respectively, knowledge, success, or gain (581).

Plato has clear views about which of these three elements ought to be the dominant one. As one would expect from his view of the Forms as the ultimate realities knowable only by intellect, it is Reason that Plato thinks ought to control both Spirit and Appetite. But each part of the soul has its proper role to play; the ideal for man is a harmonious agreement between the three elements of his soul, with Reason in control (441–2). This ideal condition Plato describes by the Greek word *dikaiosune,* which is standardly translated as 'justice'; there can be no exact English translation, but as applied to the individual person, perhaps 'well-being' or even 'mental health' conveys better the sort of concept Plato uses. Like Socrates before him and much Greek philosophy after him, Plato's emphasis is on the intellect, on knowledge. But this emphasis is simultaneously on the moral, because of his view that virtue, how to live well, is a matter for human *knowledge* rather than just conflicting opinion. There is such a thing as the truth about how we ought to live, and this truth can be known by the human intellect when we achieve knowledge of the perfect unchanging immaterial Forms.

The chief remaining feature of Plato's theory of the nature of man is that we are ineradicably social. The individual person is not self-sufficient, for he has many needs which he cannot supply for himself. Even on the level of the material needs for food, shelter, and clothing, one person can hardly supply all these things for himself with absolutely no reliance on others. Such a person would be spending most of his time in the struggle for survival, he would have little time for distinctively human activities such as friendship, play, art, and learning. Again, there is the manifest fact that different individuals have different aptitudes and interests; there are farmers, craftsmen, soldiers, administrators, etc., each fitted by nature, training, and experience to specialize in one kind of task. Such division of labour is vastly more efficient than the somewhat unrealistic alternative (369–70). According to Plato, and again this is a typical Greek view, to live in society is natural to man; anything else is less than human.

Diagnosis

The Forms define Plato's ideals for man and human society, and when he looks at the facts he finds that they are very far from these ideals. Most individual people do not manifest that harmony of the three parts of the soul which Plato calls 'justice.' Nor do human societies manifest that harmony and stability which he also calls 'justice.' Plato devotes a section of the *Republic* (543–76) to a diagnosis of the various types of imperfect society and the corresponding types of imperfect individual. In a 'timarchic' society such as that of Sparta it is the ambitious, competitive, soldierly kind of person who succeeds, and intelligence is not valued. In an 'oligarchy,' political power is in the hands of the rich, and the successful individual is the grasping money-maker. Plato took a very jaundiced view of democracy as he understood it, influenced no doubt by his experience of Athenian politics. He thought it absurd to give every person an equal say, since not everyone is equally knowledgeable about what is best for society. He criticizes the typical individual in a democratic society as lacking in discipline, living only for the pleasures of the moment. Tyranny, Plato thinks, is the typical sequel to the anarchy and chaos resulting from the unbridled liberty of democracy; one leader gains absolute power and maintains it by such unscrupulous means that the most criminal elements in human nature find their expression in the violent kind of society that results. Plato concludes that each of these types of man and society departs further from the ideal and reaches a further level of unhappiness (576–87).

The defects in human nature which Plato disagnoses are intimately related to the defects he finds in human societies. I doubt if one can attribute to him either of the simple views that individuals are to blame or that society is basically wrong. He would say, rather, that the two are interdependent. An imperfect society produces imperfect individuals, and imperfect individuals make for an imperfect society. One cannot have 'justice' in the state without having it in individuals, nor vice versa. For justice is the same thing in both cases—a harmony between the natural constituents, each doing its own job (435); and, correspondingly, injustice is disharmony. The problem, then, is how to establish harmony in individual and state.

Prescription

'There will be no end to the troubles of states, or of humanity itself, till philosophers become kings in this world, or till those we now call kings and rulers really and truly become philosophers, and political power and philosophy thus come into the same hands' (473). This is the essence of Plato's prescription, as he states it himself. He is aware that it sounds absurd, but given his theory of Forms and his theory of human nature we can see the rationale for it. If there is such a thing as the truth about how we ought to live, and if this truth can be known by those who are able and willing to learn, then those who have this knowledge are the only people who are properly qualified to direct the running of human society. Philosophers are those who have attained this knowledge by coming to know the Forms, so if society is governed by philosophers, the problems of human nature can be solved. The perfect state is that which is ruled by perfect men, and the notion of perfection here contains the intellectual and the moral and political rolled into one. So the *Republic* is at one and the same time a blueprint for the perfect state, and an analysis of the nature of that moral and intellectual knowledge which, Plato thinks, makes a perfect man.

In order to produce such perfect individuals, an elaborate system of education is necessary (376–412 and 521–41). Plato is thus the first of those who see education as the key to constructing a better society. And like many after him, he envisages various stages of education, the higher stages being open only to those qualified to undertake them, the élite who will be the future rulers of society. For these latter, the emphasis is on mathematics and philosophy, those disciplines which lead the mind to knowledge of the Forms and to a love of the truth for its own sake. The élite thus produced will prefer to philosophize more than anything else, but they will respond to the call of duty and will apply their knowledge to the running of society. After experience in subordinate offices, they will be ready for supreme power. Only such lovers of truth will be impervious to temptations to misuse their power for personal gain, for they will value the happiness of a right and rational life more than any material riches (521).

What then of the rest of society–the non-élite? There are various functions which have to be performed, and a division of labour is the natural and efficient way of organizing this. Plato makes a basically threefold division of his ideal society (412–27). Besides the philosopher-rulers, there is to be a class traditionally called the Auxiliaries, who perform the functions of soldiers, police, and civil servants. It is they who will put the directions of the Rulers into effect. The third class has no special name, but will contain the workers of all kinds–farmers, craftsmen, traders, etc., all those who produce the material necessities of life. The division between these three classes will be strict; in fact, Plato says that the 'justice' or well-being of the society depends on each person performing his own proper function and not interfering with others (432–4). Like the well-being of the individual, which Plato treats of immediately afterwards in analogous fashion, the health of society consists in a harmonious working together of its three main constituents. He says that his purpose in founding the state is not to promote the happiness of any one class, but, so far as possible, of the whole community (420). He thinks that the state will be harmonious and stable only if the strict threefold class-division is maintained, so each class must be persuaded that it is their business only to perfect themselves in their own job, and they must be content with such degree of happiness that their place in society permits (421).

Critical Discussion of Plato's Theory

The *Republic* is one of the most influential books of all times. The above sketch can only give a sample of the richness and diversity of the ideas it contains, and can be no substitute for a reading of the book itself. But I now want to go on to outline some of the main points of doubt about Plato's theory, diagnosis, and prescription, in order to start the reader thinking about them critically.

We need not enter into the many logical, epistemological, and metaphysical problems of the theory of Forms—these are still the subject of technical discussion by professional philosophers. But the moral aspect of the Forms is central to our purpose here, for we have seen that the theory that there is such a thing as the truth about how men ought to live is fundamental to Plato's treatment of the problems of human nature. Now it hardly needs saying that this assumption is a controversial one. Many people down the ages to the present day maintain that many, if not all, of the questions in morals and politics have no universally 'true' answers. It may be said that what is right varies from society to society, or that there is no one right answer even within a given society. So that if two people give different answers to moral questions then there is no truth and falsity of the matter, only a difference in taste, like one person liking beer and another preferring cider. This whole question of the objectivity of value-judgments is of course fundamental in moral philosophy, and is the subject of continuing dispute. So we must ask whether Plato has given us any adequate reason to believe that there are objective standards in morals and politics. Since he gives no direct argument for this conclusion, this must be one of the most fundamental points of doubt about his theory.

Even if there are such objective standards, Plato's theory requires also that there are rational methods to find out what they are (the education of the philosopher-rulers is supposed to teach such methods). But what if educated men sincerely disagree about fundamental questions of morals and politics—as we know very well that they often do. Is there any rational way of showing which is the right view? Plato has hardly shown that there is such a method of settling disagreements. He himself seems almost to pass beyond rationality in some places, when he talks of philosophers eventually attaining a vision of the Form of the Good itself, which will illuminate them like the blinding light of the sun (508–9). But what if in such 'vision' different philosophers claim to see different things—is there any way of determining which is right, or can there be only a conflict of opposing claims? When someone thinks he knows the ultimate truth about some important question, it is easy for him to be intolerant of anyone who disagrees, and even to feel justified in forcing his view on those who disagree (as the history of religious controversies bears witness). Plato thinks that philosophers are capable of knowing the absolute truth about how to rule society, and are therefore justified in wielding absolute power. Such a view is in striking contrast to that of Socrates, who was always conscious of how much he did *not* know, and claimed superiority to unthinking men only in that he was aware of his own ignorance where they were not.

There is much that can be questioned in Plato's theory of individual human nature. Is the soul a non-material thing? Is it immortal and indestructible? In what sense, if any, can the soul be said to have parts? And is the threefold division into

Reason, Spirit, and Appetite adequate? Only about the latter question shall I say anything more here. Threefold divisions have been popular in several theories of human nature, and perhaps Plato's will do as a first approximation, distinguishing some elements in human nature which can conflict with each other. But it is hardly a rigorous or exhaustive division, even if one relabels the parts in modern terms as intellect, personality, and bodily desire. Emotion is something that can involve all three, for instance.

There are, I think, two main criticisms to be made of Plato's blueprint for his ideal society. The first concerns his requirement that perfect men—philosopher-kings—should have absolute political power. But is there really any guarantee that any process of education, however well designed and well executed, can produce absolutely perfect men? Plato's view that philosophers will be such lovers of truth that they will never misuse their power seeme naively optimistic. Do we not need to set up a political system which will guard against the possibility of the misuse of power? Given that all men are imperfect in some way, is it not unrealistic to base a blueprint on the idea that there could be perfect men? Any realistic political system must deal with men as they are, not men as we would like them to be. Plato seems to ask himself the question 'Who is qualified to wield absolute power?' but should we not rather ask the question 'How can we ensure that nobody has absolute power?' Plato dismisses democracy rather quickly and unfairly; admittedly he is thinking of Athenian-type democracy in which every citizen would have a vote on major decisions, which would indeed be a cumbersome if not impossible system in a state of any size. But the basic idea of modern parliamentary democracies—that a government must submit itself for re-election within a certain fixed period of time—provides the kind of safety mechanism for dismissing rulers which is totally absent from Plato's *Republic*. Democracy of this kind may be inefficient and imperfect in various ways, but is not the alternative of absolute power with no guarantee against its misuse very much worse?

The second criticism is that Plato seems more concerned with the harmony and stability of the whole state than with the well-being of the individuals in it. We have already noted one place where he says something like this (420); at 519–20 he says

> The object of our legislation is not the welfare of any particular class, but of the whole community. It uses persuasion or force to unite all citizens and make them share together the benefits which each individually can confer on the community; and its purpose in fostering the attitude is not to enable everyone to please himself, but to make each man a link in the unity of the whole.

These passages can be interpreted in innocuous and in sinister ways. We are usually in favour of 'community spirit,' of each person contributing something to the well-being of the whole society, and of certain laws (e.g., against murder and theft) being enforced on all. But Plato's blueprint seems to envisage rather more than this, in his strict threefold class-division and his insistence that the harmony and stability of the state requires that each person fulfil his allotted function and that alone. The ruler must rule, even if that is not what he would really like to do, and similarly the auxiliaries must be auxiliaries, and the workers must work. This is what Plato calls 'justice' in the state, and it is not at all what we mean by the term, which implies something like equality before the law, and fair shares for all. If the worker is not content to be a worker and have no share whatsoever in politics, then Plato would, I think, forcibly compel him to remain in his station. And the advantage is not all on the side of the rulers and auxiliaries either, for they are not permitted either private property (416) or family life (457). It does seem then that Plato is prepared to deny some of what are widely thought to be essential requirements for individual happiness, in the interests of setting up a stable state which conforms to his ideals. But what is the point of a stable society unless it serves the interests of the individuals in it? Stability and efficiency are valuable, but they are certainly not the only criteria, or even perhaps the most important, for the well-being of a society.

So Plato's practical political programme of giving unrestricted power to a wise élite must be severely criticized. And his philosophical theory of Forms is subject to many philosophical objections. But his general ideas that human reason can attain knowledge through education, and that such knowledge can contribute not only to individual fulfilment but to the wise government and reform of society, are ones which almost everyone now accepts. Perhaps we do not realize how much of our heritage we owe to classical Greek thinkers in these and other respects.

In this chapter I have concentrated on one dialogue of Plato, and the reader should remember that he wrote much else, not all of it in the same vein as the *Republic,* and some of it (many scholars think) expressing the less metaphysical, more purely ethical philosophy of his great mentor Socrates.

For Further Reading

Basic text: *Republic* (many translations and editions). The translation by H. D. P. Lee in the Penguin Classics series (Penguin, London, 1955) helpfully divides up the text by subject matter. Other dialogues of Plato are also available in the same series. In the United States, see the translation with introduction and notes by F. M. Cornford (Oxford University Press paperback, New York, 1951). In Part II of *The Study of Human Nature,* edited by Leslie Stevenson (Oxford University Press, New York, 1981), there are relevant excerpts from Plato's *Republic,* along with contrasting selections from Aristotle and Lucretius.

For a general introduction to Plato's philosophy, see *The Philosophy of Plato,* by G. C. Field (Oxford University Press, London, 2nd edn. 1969). This also contains further bibliography. For a more detailed recent study, see *An Introduction to Plato's Republic,* by Julia Annas (Oxford University Press, New York, 1981).

For a hostile attack on Plato's political programme, see *The Open Society and Its Enemies,* Volume 1, by K. R. Popper (Routledge & Kegan Paul, London, 4th revd. edn. 1962; Princeton University Press paperback, Princeton, N.J., 1966). Anyone reading this will recognize the source of many of my criticisms of Plato.

4

Christianity: God's Salvation

In the introductory chapters, I suggested that Christianity contains a theory of the universe, a theory of man, a diagnosis, and a prescription; and I have already mentioned some of the standard objections. Christian doctrines have of course changed and developed over the two thousand years of their history, and the present time is perhaps a particularly confusing one, when there is wide disagreement about just what the essential doctrines are. Within the three main divisions (Roman Catholicism, Eastern Orthodoxy, and Protestantism) there are many more subdivisions and differences. Although all will acknowledge their derivation from the Old and New Testaments, and to some extent from the early Creeds and statements of the Church, there is disagreement about in what sense these sources are authoritative–some emphasize texts, others the tradition of the institutional Church, others the religious experience of individual believers. Obviously there is some difficulty in treating Christianity as a 'theory' on a par with the others in this book, for the Bible is not a text by a single author, and the religious traditions based on it are notoriously diverse. Nevertheless, they presumably have some very basic claims in common, so what I will do in this chapter is to try to sort out what these are, and to point to some of the main difficulties they face.

Theory of the Universe

First, then, let us consider the basic Christian claim about the nature of the universe, that God exists. What *sort* of God is thus asserted to exist? Not, surely, a God who is literally 'up there,' located somewhere in space and time. When the Russian cosmonauts reported that they hadn't met God in their space travels, this was surely no genuine evidence against Christianity. The Christian God is certainly not supposed to be one object among others in the universe; He does not occupy a position in space or last for a certain length of time. Neither is He to be identified with the whole universe, the sum total of everything that exists, as some writers (e.g., Spinoza) have said. This is pantheism, not Christianity. Traditionally, the Christian God is transcendent as well as immanent–although in some sense present everywhere and all the time, He is also beyond or outside the world of things in space and time (Psalm 90:2, Romans 1:20). Some contemporary theologians seem prepared to deny this, and to define God as ultimate reality, the ground of all being, or as that which concerns us ultimately; but such definitions seem to be quite compatible with what was traditionally called atheism! We even hear of people calling themselves 'Christian Atheists.' It does appear that in the effort to accommodate Christianity to the modern mind, such doctrines really deny what they are trying to defend.

The transcendent existence of God is, then, essential to Christianity. But there are genuine philosophical difficulties about the doctrine. It was once widely supposed that there were valid arguments to prove the existence of God, but Hume and Kant destructively criticized those arguments in the eighteenth century. Some Christians (mostly Roman Catholics) have continued to claim to be able to prove the existence of God, but the validity of their arguments is of course hotly disputed by non-believers. Many Christians will now agree that God's existence can be neither proved nor disproved by reason alone, that belief in Him is a matter for faith rather than argument.

But still, *what* is it that one believes when one believes in

God? This is where the crucial modern debate about meaningfulness and verifiability begins. If God is transcendent, He is of course not visible or tangible, or observable by any of the methods of science. But He is not a mere abstraction like numbers and the other objects of mathematics, for He is supposed to be a personal Being who loves us. If, then, neither empirical observation nor purely logical reasoning can count for or against His existence, just what is being asserted by the believer in God? In Chapter 2 we noted how the suffering and evil in the world would seem to be evidence against there being an omniscient, omnipotent, and benevolent God, and yet the Christian does not necessarily count this as telling against his belief. He may say that out of suffering greater good can come in the end, or that the possibility of evil must be there if men are to be genuinely free to make moral choices. But the non-believer may still wonder why God could not have made the world such that suffering was not the only way to good, and such that men would freely have chosen rightly. So it does look as if the Christian does not take his belief in God to be falsifiable by evidence about the actual state of the world.

Another vital part of the Christian doctrine of God is that He created the world (Genesis 1:1, Job 38:4). (This presupposes His transcendence, for the world could hardly be created by a part, or even the whole, of itself.) But it is a misinterpretation of this doctrine to say that it implies that the Creation was an event in time. Modern theologians are not necessarily dismayed by cosmological theories which imply that the universe has no beginning in time. And it is now widely accepted that the story of the creation of man in Genesis is myth (symbolic of deep religious truths) rather than history, so there is no incompatibility with the theory of evolution, despite the disputes still raised on the subject (see Chapters 9 and 10). Any Christian who still asserts the historical existence of Adam and Eve is insisting on an over-literal interpretation of the words of Scripture. But the question remains, just what *is* meant by saying that God is the creator of the world, and of man? It seems to imply that if God did not exist, the world would not exist; and that the world is somehow fundamentally in accordance with God's purpose, that there is nothing which exists save by His design or at least by His permission. It was once common to argue that the world, especially the world of living things, is very much as if it had been designed by a very intelligent and powerful Creator. But Hume and Kant effectively destroyed this 'Argument from Design'; and modern biological science has provided convincing alternative explanations of the marvellous adaptation of living systems to their environment. So theologians nowadays are much less inclined to test the doctrine of God's creation of the world by observation of the state of the world. But this raises again the question of what *sort* of statement the Christian doctrine of Creation is.

According to the verification principle, mentioned in Chapter 2, if a statement is neither verifiable by observation nor provable by logic alone then it is not literally meaningful, it cannot assert anything about what is the case, but is at best a poetic use of language, an expression of attitudes or emotions. Now some Christians have been content to say that all they are doing when they say that God exists is to affirm an attitude, perhaps that love is the most important thing in the universe, or that we should behave *as if* the universe were ruled by a loving God. But an atheist might also be willing to hold such attitudes, while still disagreeing about whether God actually does exist. Any belief which is to deserve the name of Christianity must be doing *more* than merely expressing an attitude, vital though attitudes and actions are.

Other Christians accept the challenge of the verifiability criterion of meaningfulness, and try to meet it by suggesting that in certain human experiences–moral or religious or mystical–there is the possibility of empirical verification of God. But it will inevitably be highly controversial how to describe such experiences, and the non-believer will naturally find great difficulty in interpreting any human experience in terms of a transcendent God. Another suggestion has been that in the life after death we shall be able to verify the existence and nature of God, by something like observation. But this is to meet one verifiability problem by posing another, for how can we *now* verify, or find evidence for, the reality of life after death?

Theologians who are aware of more recent work in philosophy will question the verification principle itself as an adequate criterion of meaningfulness. But they will still have to reckon with the philosophically more acceptable principle that any *factual* or *scientific* statement must be falsifiable. If the assertion of God's existence is such that no conceivable evidence could count against it, then it is hard to see how it can be an assertion of what is the case about the universe. Most believers will agree that their belief is not a scientific one, and many are attracted to the idea of science and religion giving not rival but *complementary* accounts of the universe, describing the same ultimate reality from different sides, as it were. However this still does not explain how religious statements can give any kind of knowledge of reality, if they are not in principle falsifiable. This remains as one of the most basic philosophical problems about religion, and this is why so much of the contemporary discussion in the philosophy of religion centres around questions of meaning and epistemology for religious assertions. In this book we cannot pursue these questions further, but must concentrate on Christian doctrine of human nature.

Theory of Man

The Christian doctrine of man sees him primarily in relation to God, who has created him to occupy a special position in the universe. Man is made in the image of God, to have dominion over the rest of creation (Genesis 1:26); he is unique in that he has in him something of the self-consciousness and ability to love freely which is characteristic of God Himself. God created man for fellowship with Himself, so man fulfils the purpose of his life only when he loves and serves his creator.

But although man is thus seen as fundamentally distinct from the rest of creation, he is at the same time continuous with it (if this is not a contradiction!). He is made of 'dust from the ground' (Genesis 2:7), that is, of material stuff. It is a common and recurrent misinterpretation of Christian doctrine that it asserts a dualism between the material body and an immaterial soul or mind. Such dualism is a Greek idea (we have noticed it in Plato in Chapter 3), and is not to be found in the Old or New Testaments. In the early centuries of the Church, Christian theology began to employ ideas of Greek philosophy in its formulations of doctrine, and the theory of the immaterial soul did find its way into Christian thinking and has tended to stay there ever since. Christianity is of course committed to the idea of life after death, but it is heterodox to think of this as the survival of an immaterial soul after the death of the material body. The Creeds explicitly state belief in the resurrection of the *body,* and the scriptural warrant for this is in 1 Corinthians 15:35 ff., where St. Paul says that we die as physical bodies but are raised as spiritual bodies. Admittedly, it is not clear what a spiritual body is, but St. Paul does use the Greek word *soma,* which means 'body.'

This belief in life after death by resurrection of the body is, I think, another of the essential doctrines of Christianity. To interpret the doctrine just as 'the evil that men do lives after them,' or to take the promise of eternal life (John 4:14) as only of a new way of life in this world, is to evacuate the doctrine of one of its essential contents. The humanist can join with the Christian in seeking a regeneration of man as we know him, an escape from selfishness and pride; it is the hope of a survival of the individual person into the eternal dimension which is distinctively Christian. But as before, this essential transcendent element in the Christian claim runs into philosophical difficulties. If bodies are resurrected, presumably, being *bodies* of some kind, they have to occupy space and time of some kind. Now it is surely not meant that they exist somewhere in the space in which we are located–no Christian should expect a spaceman to be able to come across the resurrected bodies of St. Paul, Napoleon, or Auntie Agatha! So it seems that what we have to try to make sense of is the idea that there is a space in which resurrected bodies exist which has no spatial relations with the space in which we exist. The question of time is at least as difficult. Presumably it is not necessarily intended that there is some time in the future at which the resurrection will take place (although when Paul says 'we shall all be changed, in a moment, in the twinkling of an eye, at the last trumpet' (1 Corinthians 15:51–2) it does sound rather like this). Is there then a time system which has no temporal relation to us, or are the resurrected bodies timeless, in which case what sense can be made of the idea of resurrected *life?* (For life, as we understand it, is a process in time.)

It is another misinterpretation of the Christian doctrine of man to identify the distinction between good and evil with that between spirit and body, or mind and matter. This view that all matter is basically evil is not a Christian one, even if it had its influence on early Christian thought. St. Paul's distinction between spirit and flesh (Romans 8) is not between mind and matter, but between regenerate and unregenerate man. We shall look at the idea of regeneration in a moment.

The most crucial point in the Christian understanding of human nature is the notion of freedom, the ability to love, which is the image of God Himself. Plato (and Greek philosophy in general) puts the emphasis on the intellect, on the ability of man to attain knowledge of theoretical and moral truth; the true purpose of human life is thought to be attainable only by such as are able to gain such knowledge. Christianity, in contrast, puts the emphasis, not just on morality or virtuous living, but on the foundation in character and personality from which such life proceeds. The attainment of the true purpose of human life—love of God, and life according to His will—is open to all regardless of intellectual ability (1 Corinthians 1:20). 'If I understand all mysteries and all knowledge . . . but have not love, I am nothing' (1 Corinthians 13:2). This love (for which the Greek word was *agape,* formerly translated as 'charity') is not to be identified with merely human affection of any kind, it is ultimately divine in nature, and can be given only by God.

Diagnosis

Given the Christian doctrine of man as made by God, the Christian diagnosis of what is basically wrong with man follows easily. He has sinned, he has misused his God-given free will, he has chosen evil rather than good, and has therefore disrupted his relationship to God (Isaiah 59:2).

But again this doctrine of the fall of man needs disentangling from misinterpretation. The Fall is not a particular historical event–the Genesis story of Adam and Eve, the snake and the apple, is myth rather than historical narrative. It is a symbol of the fact that all men are subject to sin, that there is a fatal flaw in our very nature. But this doctrine of 'original sin' does not imply that we are totally and utterly depraved, that we can do nothing good. It is that nothing we can do can be perfect by God's standards: 'All have sinned, and fall short of the glory of God' (Romans 3:23). Sin is *not* basically sexual in nature, although ever since St. Augustine there has been a tendency in Christian thought to identify sin with the lusts of the flesh. Sex has its rightful place within Christian marriage; the true nature of sin is nothing essentially bodily, but rather the assertion of man's will against God's and his consequent alienation from God.

The Fall of man somehow involves the whole creation in evil (Romans 8:22); everything is in some way 'short of the glory of God.' But it is not necessary for Christians to postulate some kind of personal Devil to express this idea of cosmic Fall. And it is heresy to believe in twin and equal powers of good and evil; for the Christian, God is creator of all, and is ultimately in control of all. But this belief runs directly into the problem of evil, which we have already noted.

Prescription

The Christian prescription for man is based on God, just as much as the theory and diagnosis is. If God has made man for fellowship with Himself, and if man has turned away and broken his relationship to God, then only God can forgive man and restore the relationship. Hence the typically Biblical idea of salvation, of a regeneration of man made possible by the mercy, forgiveness, and love of God. In the Old Testament there is the covenant made between God and His chosen people (Exodus 19:5), by which God redeems them from their bondage in Egypt and promises that they will be His people if they keep His commandments. When the Jews fail to obey

God's laws, there comes the idea of God using the events of history, such as defeat by neighbouring nations, to chastise them for their sin (a theme which recurs throughout the histories and prophets in the Old Testament). And then there is the idea of God's merciful forgiveness, His blotting out of man's transgressions, and His regeneration of man and the whole of creation (Isaiah chapters 43–66).

But it is in the New Testament, in the life and work of Jesus, that we find the distinctively Christian (rather than Jewish) idea of salvation. The central claim is that God was uniquely present in the particular human being Jesus, and that God uses his life, death, and rising again to restore men to a right relationship with Himself. No belief can properly call itself Christian unless it accepts the essential content of these claims. It is not enough to say that Jesus was a great man, a man of genius, or even a man of supreme religious genius above all others before or since. The Christian claim is traditionally expressed in the doctrine that Jesus is the Son of God, both human and divine, the eternal Word made Flesh (John 1:1–18). The early philosophical versions of this doctrine–two natures in one substance, and so on–are perhaps not essential. But the basic idea of incarnation, that God is *uniquely* present in Jesus, is. And equally essential is the idea of atonement, that the particular historical events of the life, death, and resurrection of Jesus (and their continual representation by the Christian Church) are the means by which God reconciles His creation to Himself. It is not enough to say that Jesus' life and death are an example to us all. It is implied that the resurrection of Jesus really happened (1 Corinthians 15:17), however flagrant the contradiction with all the known laws of nature. (The idea of the Virgin Birth is just as improbable, but perhaps less important.)

These doctrines of incarnation and atonement defy human rationality, and indeed their formulation has provoked much disagreement within Christianity. How can a particular person be a member of the transcendent Godhead? The Christian doctrine of the Trinity–that there are three persons in one God (Father, Son, and Holy Spirit)–multiplies the conceptual problems rather than solves them. The standard thing to say, of course, is that these are mysteries rather than contradictions, that human reason cannot expect to be able to understand the infinite mysteries of God, that we only believe in faith what God has revealed of Himself to us. But the trouble with this kind of statement is that it can appeal only to those already disposed to believe, it can do nothing to answer the genuine conceptual difficulties of the sceptic. The same applies to atonement: not many Christians will interpret this like the propitiatory sacrifices of the Old Testament, as if God requires blood to be shed (any blood, even that of the innocent) before he will forgive sins, but it is still an enormous mystery how the crucifixion of a Jewish religious teacher at the hands of the Roman governor Pontius Pilate somewhere around A.D. 30 can effect a redemption of the whole world from sin.

The Christian prescription is not quite complete, however, with the saving work of Jesus Christ. It remains for this salvation to be accepted and made effective in each individual person, and to be spread throughout the world by the Christian Church. Each person must accept the redemption that God has effected for him in Christ, and become a member of the Church, the community in which God's grace is active. Different Christian traditions have emphasized individual acceptance or Church membership, respectively, but all will agree that both are necessary. Thus the regeneration of man and world takes effect: 'if anyone is in Christ, he is a new creation' (2 Corinthians 5:17). There is not necessarily a single experience of conversion in each individual, nor does regeneration take place all at once; it must be a lifelong process, which looks beyond this life to the resurrection of the body for its completion and perfection (Philippians 3:12).

A final conceptual problem (or mystery) arises over the parts played by man and God in the drama of salvation. The fundamental Christian conception is certainly that redemption can only come from God, through His offering of Himself in Christ. If we are saved, we are saved by this free grace of God, not by anything that we can do ourselves (Ephesians 2:8). Yet, just as clearly, the Christian doctrine is that man's will is free; it is by his own choice that he sinned in the first place, and it must be by his own choice that he accepts God's salvation and works out its regeneration in his life. The New Testament is full of exhortations to repent and believe (Acts 3:19), and to live the life that God makes possible through the regenerating power of the Holy Spirit (Galatians 5:16). There is thus a tension, if not a contradiction, between the insistence that all is due to God and the exhortation that salvation depends on man's response. St. Augustine emphasized the former, and Pelagius the latter; in this controversy the problem of the freedom of the will arises as a crucial internal problem for Christian theology. Although Pelagius was condemned as heretical, the doctrine of human free will must still remain as an essential element in Christian belief, difficult as it is to reconcile with the theory of the complete sovereignty of God.

Many thinking Christians would acknowledge that there are all these conceptual problems in the essential Christian doctrines. But they would emphasize that Christianity is more than a theory, it is a way of life; and though it may be called an ideology, it is not a political ideology like Marxism. They remain practising Christians, and accept the basic theory despite its difficulties, because of what they find in the life and worship of the Church: a certain growth in the inner or 'spiritual' life which they do not find elsewhere. There can be no complete assessment of Christianity unless this is considered.

For Further Reading

Basic text: The Bible (many versions and translations); I have been quoting from the Revised Standard Version. A commentary such as *Peake's Commentary on the Bible* will help to elucidate many difficulties. So will *The New Oxford Annotated Bible with the Apocrypha: Revised Standard Version.* An Ecumenical Study Bible, edited by Herbert G. May and Bruce M. Metzger (Oxford University Press, New York, 1974).

The Existence of God, edited by John Hick (Collier–Macmillan, London, 1964; Macmillan paperback, New York, 1964), is a collection of readings from classic authors on the traditional arguments for and against the existence of God, together with some readings from modern authors on the verifiability question.

Philosophy of Religion, by John Hick (Prentice-Hall, Englewood Cliffs, N.J., 3rd edn. 1983, paperback), is an introductory book which concentrates attention on the Judaic–Christian concept of God and admirably surveys the contemporary philosophical discussions of it. *Religion,* by Leszek Kolakowski (Fontana, London, 1982), is a subtle study of the uneasy relationship between reason and religion.

For more on Christian understandings of human nature, see Reinhold Niebuhr's classic Gifford lectures delivered in 1939 and collected in *The Nature and Destiny of Man* (Charles Scribner's Sons, New York, 1964); also see E. L. Mascall, *The Importance of Being Human* (Columbia University Press, New York, 1958), which clearly presents a neo-Thomist view; and J. Macquarrie, *In Search of Humanity* (SCM Press, London, 1982; Crossroad, New York, 1983), which is much influenced by existentialist philosophy.

5

Marx: Communist Revolution

In comparing Marxism to Christianity in the introductory chapters I have already sketched some of the main ideas of Marxism and some of the hackneyed objections to them. In this chapter I would like to go a little deeper by giving an introduction to Marx's life and work, followed by a critical analysis of his theory of history, theory of man, diagnosis, and prescription. I shall not attempt to define or discuss the many subsequent varieties of Marxism and communism; I concentrate on the ideas of Karl Marx himself. (Although Marx and Engels wrote some works in collaboration, there is no doubt that Engels's contribution was relatively minor.)

Life and Work

Karl Marx was born in 1818 in the German Rhineland, of a Jewish family who became Christian; he was brought up as a Protestant, but soon abandoned religion. He displayed his intellectual ability early, and in 1836 he entered the University of Berlin as a student in the faculty of Law. The dominant intellectual influence in Germany at that time was the philosophy of Hegel, and Marx very soon became immersed in reading and discussing Hegel's ideas, so much so that he abandoned his legal studies and devoted himself completely to philosophy. The leading idea in Hegel's work was that of historical development. He held that each period in the history of each culture or nation has a character of its own, as a stage in the development from what proceeded it to what will succeed it. Such development, according to Hegel, proceeds by laws which are fundamentally mental or spiritual; a culture or nation has a kind of personality of its own, and its development is to be explained in terms of its own character. Hegel took this personification even further and applied it to the whole world. He identified the whole of reality with what he called 'the Absolute,' or world-self, or God (this is of course a pantheist rather than a Christian concept of God), and interpreted the whole of human history as the progressive self-realization of this Absolute Spirit. 'Self-realization' is thus seen as the fundamental spiritual progress behind all history. It is the overcoming of what Hegel called 'alienation,' in which the knowing person (the subject) is confronted with something other than or alien to himself (an object); somehow this distinction between subject and object is to be merged in the process of Spirit realizing itself in the world.

The followers of Hegel split into two camps over the question of how his ideas applied to politics. The 'Right' Hegelians held that the process of historical development automatically led to the best possible results. So they saw the contemporary Prussian State as the ideal culmination of preceding history. Accordingly, they held conservative political views, and tended to emphasize the religious elements in Hegel's thought. The 'Left' or 'Young' Hegelians thought that the ideal had yet to be realized, that the nation-states of the time were very far from ideal, and that it was the duty of men to help change the old order and assist the development of the next stage of human history. Accordingly, they held radical political views, and tended to identify God with man, thus taking a fundamentally atheist view. One of the most important thinkers in this direction was Feuerbach, whose *Essence of Christianity* was published in 1841. Feuerbach held that Hegel had got everything upside-down, that far from God progressively realizing Himself in history, the situation is really that the ideas of religion are produced by men as a pale reflection of this world, which is the only reality. It is because men are dissatisfied or 'alienated' in their practical life that they need to believe in illusory ideas. Accordingly, metaphysics is just 'esoteric psychology,' the expression of feelings within ourselves rather than truths about the universe. Religion is the expression of alienation, from which men must be freed by realizing their purely human destiny in this world. Feuerbach is, then, one of the most important sources of humanist thought.

This was the intellectual atmosphere of Marx's formative years. His reading of Feuerbach broke the spell that Hegel had cast on him, but what remained was the idea that in Hegel's writings the truth about human nature and society was concealed in a kind of inverted form. As we shall see, the notions of historical development and of alienation play a crucial role in Marx's thought. He wrote a critique of Hegel's *Philosophy of Right* in 1842–3, and at the same time became editor of a radical journal of politics and economics called the *Rheinische Zeitung*. This journal was soon suppressed by the Prussian government, so Marx emigrated to Paris in 1843. In the next two years there he encountered the other great intellectual influences of his life, and began to formulate his own distinctive theories. His wide reading included the British economist Adam Smith and the French socialist Saint-Simon. He met other socialist and communist thinkers such as Proudhon, Bakunin, and Engels. (This was the beginning of his lifelong friendship and collaboration with Engels.) In 1845 he was expelled from Paris, and he moved to Brussels.

In these years in Paris and Brussels Marx formulated his so-called 'Materialist Theory of History.' By inverting Hegel's view as Feuerbach had suggested, Marx came to see the driving force of historical change as not spiritual but material in character. Not in men's *ideas,* and certainly not in any sort of national or cosmic personality, but in the *economic* conditions of men's life, lay the key to all history. Alienation is neither metaphysical nor religious, but really social and economic. Under the capitalist system labour is something external and alien to the labourer; he does not work for himself but for someone else–the capitalist–who owns the product as private property. This diagnosis of alienation is to be found in the 'Economic and Philosophical Manuscripts' which Marx wrote in Paris in 1844, but which did not become generally available in English until the 1950s. The materialist conception of history is to be found in other works of this period—*The Holy Family* of 1845, *The German Ideology* of 1846 (written with Engels), and *The Poverty of Philosophy* of 1847.

In Brussels Marx became involved with the practical organization of the socialist and communist movement, a task which occupied much of the rest of his life. For he saw the main purpose of his work as 'not just to interpret the world, but to change it' (as he put it in his *Theses on Feuerbach* in 1845). Convinced that history was moving towards the revolution by which capitalism would give way to communism, he tried to educate and organize the 'proletariat'—the class of workers to whom he thought victory would go in the imminent struggle. He was commissioned to write a definitive statement of the aims of the international communist movement, and together with Engels, he produced the famous *Manifesto of the Communist Party,* which was published early in 1848. Soon afterwards in that year (although hardly as a result of the *Manifesto!*) there were abortive revolutions in several of the major European countries. After their failure Marx found himself expelled from Belgium, France, and Germany, so in 1849 he went into exile in London, where he remained for the rest of his life.

In London Marx endured a life of poverty, existing on occasional journalism and gifts from Engels. He began daily research in the British Museum and continued to organize the international communist movement. In 1857–8 he wrote another series of manuscripts called *Grundrisse,* sketching a plan of his total theory of history and society. Not until 1973 has the complete text of these been available in English. In 1859 he published his *Critique of Political Economy,* and in 1867 the first volume of his most substantial work, *Das Kapital.* These last two works contain much detailed economic and social history, reflecting the results of Marx's labours in the British Museum. Although there is less evidence of Hegelian philosophical ideas such as alienation, Marx was still trying to apply his materialist interpretation of history to prove the inevitability of the downfall of capitalism.

It is these later works, from the *Communist Manifesto* onwards, that have been best known and have formed the basis of much communist theory and practice. In them we find German philosophy, French socialism, and British political economy, the three main influences on Marx, integrated into an all-embracing theory of history, economics, and politics. This is what Engels came to call 'scientific socialism'; for Marx and Engels thought they had discovered the correct *scientific* method for the study of history, and hence the truth about the present and future development of the society of their time. But the recent publication of the earlier works, particularly the Paris Manuscripts of 1844, has shown us much about the origin of Marx's thought in Hegelian philosophy, and has revealed the more philosophical nature of his early ideas. So the question has been raised whether there were two distinct periods in his thought—an early phase which has been called humanist or even existentialist, giving way to the later and more austere 'scientific socialism.' I think it is fair to say that the consensus of opinion is that there is a continuity between the two phases, that the theme of alienation is buried but still there in the later work; the contents of the *Grundrisse* of 1857–8 seem to confirm this. My discussion of Marx will therefore be based on the assumption that his thought is not discontinuous. My page references in what follows are to the Pelican book *Karl Marx: Selected Writings in Sociology and Social Philosophy,* which is perhaps the most useful of the many volumes of selected readings from Marx and Engels, containing as it does selections from both the early and late phases. (Page references to the American paperback edition are supplied at the end of the chapter.)

Theory of the Universe

Let us now begin our critical analysis of Marx's main theory. He was of course an atheist, but this is not peculiar to him.

What is distinctive of his understanding of the world as a whole is his interpretation of history. He claimed to have found the *scientific* method for studying the history of human societies, and looked forward to the day when there would be a single science, including the science of man along with natural science (p. 85). Accordingly, he held that there are universal *laws* behind historical change, and that the future large-scale course of history can be *predicted* from knowledge of these laws (just as astronomy predicts eclipses). In the preface to the first edition of *Capital,* Marx compared his method to that of the physicist and said 'the ultimate aim of this work is to lay bare the economic law of motion of modern society'; he also talked of the natural laws of capitalist production 'working with iron necessity towards inevitable results.' He agreed with Hegel that each period in each culture has a character of its own, so that the only truly universal laws in history could be those concerned with the processes of *development* by which one stage gives rise to the next. He divided history roughly into the Asiatic, the ancient, the feudal, and the 'bourgeois' or capitalist phases, and held that each had to give way to the next when conditions were ripe (p. 68). Capitalism was expected to give way, just as inevitably, to communism (pp. 150–1).

However, there are strong reasons for questioning the concept of laws of history. Certainly, history is an *empirical* study in that its propositions can and must be tested by evidence of what has actually happened. But it does not follow that it has the other main feature of a *science,* that it tries to arrive at *laws,* that is, generalizations of unrestricted universality. For history is after all the study of what has happened on one particular planet in a finite period of time. The subject matter is wide, but it is one *particular* series of events; we know of no similar series of events elsewhere in the universe, so human history is unique. Now for any particular series of events, even an apparently simple one like the fall of an apple from a tree, there is no clear limit to the number of different scientific laws that may be involved–the laws of gravity and mechanics, of wind pressure, of elasticity of twigs, of decay of wood, etc. If there is no one law governing the fall of an apple, then how much more implausible it is to postulate a general law of development behind the whole of human history.

The idea that the course of history is predetermined, so that one main function of historical study is large-scale prophecy, is at least as questionable. Certainly there may be some long-term and large-scale *trends* to be found, for instance the increase of human population since the Middle Ages. But a trend is not a *law;* its continuation is not inevitable but may depend on conditions which can change. (It is obvious that population cannot increase indefinitely, indeed its growth might be quite suddenly reversed by nuclear war or widespread famine.)

The other main feature of Marx's view of history is what is called his materialist conception of history. This is the theory that the supposed laws of history are *economic* in nature, that 'the mode of production of material life determines the general character of the social, political, and spiritual processes of life' (p. 67, cf. pp. 70, 90, 111–12, etc.). The economic structure is supposed to be the real basis by which everything else about a society is determined. Now it is undeniable that economic factors are hugely important, and that no serious study of history or social science can ignore them. Marx can take some of the credit for the fact that we now recognize this so readily. But he himself is committed to the more dubious assertion that the economic structure of a society *determines* its 'superstructures.'

This proposition is difficult to interpret, for it is not clear where the dividing line between basis and superstructure should run. Marx talks of 'the material powers of production' (p. 67) which presumably would cover land and mineral resources, tools and machines, plus perhaps the knowledge and skills of men. But he also talks of the economic structure as including 'relations of production,' which presumably means the way in which work is organized (e.g., division of labour and certain hierarchies of authority); yet the description of such organization must surely use concepts like property and money, which seem to be the kind of legal concepts that Marx would wish to put into the superstructure. If the basis includes only the material powers of production, then Marx is committed to a rather implausible 'technological determinism'; but if it includes also the relations of production, then the distinction between basis and superstructure is blurred.

From his general theory of history Marx derived a very specific prediction about the future of capitalism. He confidently expected that it would become more and more unstable economically, that the class struggle between bourgeois and proletariat would increase, with the proletariat getting both poorer and larger in number, until in a major social revolution the workers would take power and institute the new communist phase of history (pp. 79–80, 147–52, 194, 207, 236–8). Now the huge and simple fact is that this has not happened in the main capitalist countries–Britain, France, Germany, and the United States. On the contrary, the economic system of capitalism has become more stable, conditions of life for most people have improved vastly on what they were in Marx's time, and class-divisions have been blurred rather than intensified (consider the large numbers of 'white-collar' workers–office staff, civil servants, teachers, etc., who are neither industrial labourers nor industrial owners). Where communist revolutions *have* taken place, they were in countries which had little or no capitalist development at the time–Russia in 1917, Yugoslavia in 1945, China in 1949. This must surely constitute the major falsification of Marx's theory. It cannot really be explained away by saying that the proletariat have been 'bought off' by concessions of higher wages–for Marx predicted their lot would get worse. Nor is it plausible to say that colonies have formed the proletariat vis-à-vis the industrialized countries–for some, such as Scandinavia, have had no colonies, and even in the colonies conditions did improve, however slightly. To maintain Marx's theory as he stated it, in the face of such counter-evidence, makes it into a matter of blind faith, a closed system, rather than the scientific theory he claimed it to be.

Theory of Man

Except perhaps when he read Hegel's philosophy as a young man, Marx was not interested in questions of 'pure' or academic philosophy, which he would dismiss as mere speculation compared to the vital task of changing the world (p. 82). So when he is called a materialist, this refers to his materialist theory of history and not to a theory about the relation of mind to body. Certainly, he would dismiss belief in life after death as one of the illusory ideas of religion, and would emphasize that everything about the individual person (including his consciousness) is determined by the material conditions of his life (pp. 69, 85). But this could well be an 'epiphenomenalist' view—that consciousness is something non-material but entirely determined by material events–rather than a strictly materialist view that consciousness is itself material.

His view on the metaphysical question of determinism is rather ambiguous too. Of course his general view sounds determinist, with his theory of the inevitable progress of history through economic stages and his referring of all change to economic causes. And yet, just as with the Augustinian–Pelagian controversy within Christianity, there seems to be an irreducible element of free will too. For Marxists constantly appeal to their readers and hearers to realize the direction in which history is moving, and to *act* accordingly–to help bring about the communist revolution. Within Marxism there has been controversy between those who emphasize the need to wait for the appropriate stage of historical development before expecting the revolution, and those who emphasize the need to act to bring it about. But perhaps there is no ultimate contradiction here, for Marx can say that although the revolution will inevitably occur sooner or later, it is possible for individuals and groups to assist its coming and ease its birth pangs, acting as the midwives of history. Further inquiry into determinism and free will would probably be condemned as useless speculation.

What is most distinctive of Marx's concept of man is his view of our essentially *social* nature: 'the real nature of man is the totality of social relations' (p. 83). Apart from a few obvious biological facts such as the need to eat, Marx would tend to say that there is no such thing as individual human nature–what is true (and even universally true) of men in one society or period is not necessarily true of them in another place or time. Whatever a person does is an essentially social act, which presupposes the existence of other people standing in certain relations to him (pp. 91–2, 251). Even the ways in which we eat, sleep, copulate, and defecate are socially learned. This is true above all of every activity of production, for the production of our means of subsistence is typically a social activity in that it requires the cooperation of men in some way or other (p. 77). It is not that society is an abstract entity which affects the individual (p. 91), but rather that what kind of individual one is and what kind of things one does are determined by what kind of society one lives in. What seems instinctual in one society–for example, a certain role for women–may be quite different in another society. In one of Marx's typical aphorisms: 'it is not the consciousness of men that determines their being, but, on the contrary, their social being determines their consciousness' (p. 67). In modern terms, we can summarize this crucial point by saying that sociology is not reducible to psychology, that is, it is not the case that everything about men can be explained in terms of facts about individuals; the kind of society they live in must be considered too. This methodological point is one of Marx's most distinctive contributions, and one of the most widely accepted. For this reason alone, he must be recognized as one of the founding fathers of sociology. And the *method* can of course be accepted whether or not one agrees with the particular *conclusions* Marx came to about economics and politics.

But there does seem to be at least one universal generalization that Marx is prepared to make about human nature. This is that man is an *active,* productive being, who distinguishes himself from the other animals by the fact that he *produces* his means of subsistence (p. 69). It is natural for men to work for their living. No doubt there is an empirical truth here, but it seems that Marx also draws a value-judgement out of this, namely that the kind of life which is *right* for men is one of productive activity. As we shall see, this is implicit in his diagnosis of alienation as a lack of fulfilment in industrial labour (p. 177), and in his prescription for future communist society in which everyone can be free to cultivate their own talents in every direction (p. 253). No doubt it is because of this point, which is clearest in his early writings, that Marx has been called a humanist.

Diagnosis

Marx's theory of what is wrong with man and society involves his concept of alienation, which, as we have seen, is a descendant of the concept used by Hegel and Feuerbach. For Marx, alienation sums up what is wrong with capitalism; the concept rolls up together both a description of certain features of capitalist society and a value-judgement that they are fundamentally wrong. But the trouble with the notion of alienation is that it is so vague that we hardly know *which* feature of capitalism Marx is condemning. Logically, alienation is a relation, that is, it must be *from* somebody or something; one cannot just be alienated any more than one can kill without killing something. Marx says that alienation is from man himself and from Nature (p. 177). But this does not help us very much, for it is not clear how one can be alienated from oneself; and the concept of Nature involved here has obscure Hegelian roots in the distinction between subject and supposedly alien object. For Marx, Nature means the man-created world, so we can take him as saying that men are not what they should be because they are alienated from the objects and social relations that they create. The general idea that emerges from this rather mystifying terminology is that capitalist society is in some respects not in accordance with basic human nature. But it still remains to be seen what those respects are.

Sometimes it seems that private property is what Marx primarily blames for alienation, for he says that the abolition of private property is the abolition of alienation (p. 250). But elsewhere he says that 'although private property appears to be the basis and cause of alienated labour, it is rather a consequence of the latter' (p. 176). He describes this alienation of labour as consisting in the fact that the work is not part of the worker's nature, he does not fulfil himself in his work, but feels miserable, physically exhausted, and mentally debased. His work is forced on him as a means for satisfying other needs, and at work he does not belong to himself but to another person. Even the objects he produces are alien to him, because they are owned by someone else (pp. 177–8). Sometimes Marx seems to be blaming alienation on the institution of money, as a means of exchange which reduces social relationships to a common commercial denominator (pp. 179–81). Elsewhere he says that the division of labour makes man's work into an alien power opposed to him, preventing him from switching from one activity to another at will (which Marx improbably alleges will be possible in communist society; pp. 110–11). And in another passage, Marx locates the basis of social evils and the general explanation of them in the principle of the State itself (p. 223).

What then *is* Marx diagnosing as the basic cause of alienation? It may be hard to believe that anyone would seriously advocate the abolition of money (a return to a system of barter?), the disappearance of all specialization in work, or the nationalization of all property (even tooth-brushes, shirts, books, etc.?). It is the private ownership of industry–the means of production and exchange–that is usually taken as the defining feature of capitalism. And the main points in the programme of the *Communist Manifesto* are the nationalization of land, factories, transport, and banks. But it is not at all clear that such institutional changes could cure the alienation of labour which Marx describes in such psychological terms (in the early works referred to in my previous paragraph). And if the State is the basis of social evils, nationalization would make things worse, by increasing the power of the State.

It looks as if we must understand Marx as saying, at least in his early phase, that alienation consists in a lack of *community,* so that people cannot see their work as contributing to a group of which they are members, since the State is not a real community (p. 226). Such a diagnosis would suggest a prescription not of nationalization but of decentralization into genuine communities or 'communes' (in which the abolition of money, specialization, and private property might begin to look more realistic).

If this is contentious, there is a more general diagnosis implicit in Marx, which would perhaps command universal assent. This is that it is always wrong to treat any human being as only a means to an economic end. This is just what did happen in the unrestrained capitalism of the early nineteenth century, when children worked long hours in filthy conditions and died early deaths after miserable lives. Industry is made for man, not man for industry–and 'man' here must mean *all* the human beings involved. But it is of course more difficult to agree on how to give effect to this very general value-judgement.

Prescription

'If man is formed by circumstances, these circumstances must be humanly formed' (p. 249). If alienation is a social problem caused by the nature of the capitalist economic system, then the solution is to abolish that system and replace it by a better one. And we have already seen that Marx thinks that this is bound to happen anyway, for capitalism will burst asunder because of its inner contradictions, and the communist revolution will usher in the new order of things in which alienation will disappear and man will be regenerated in his true nature. Just as Christianity claims that salvation has already been enacted for us, so Marx claims that the resolution of the problems of capitalism is already on the way in the movement of history.

But Marx holds that only a complete revolution of the economic system will do. There is no point in trying to achieve limited reform such as higher wages, shorter hours, etc., for these do not alter the evil nature of the basic system, and only distract attention from the real task, which is to overthrow it. Hence the radical difference between the programme of the Communist Party and that of most trade unions and social democratic parties. This doctrine of 'the impotence of politics' follows from Marx's premises in his materialist theory of history–for if all legal and political institutions are really determined by the underlying economic system, then they cannot be used to change the economic system. However this doctrine flies in the face of the facts of the development of capitalism since Marx's time. Legal and other institutions *have* modified the economic system of capitalism very considerably, beginning with the Factory Acts of the nineteenth century which limited the worst excesses of exploitation of workers, continuing with National Insurance, unemployment benefit, National Health Services, and steady progress by trade unions in increasing real wages and decreasing working hours. In fact, many of the specific measures proposed in the *Communist Manifesto* have come into effect in the so-called capitalist countries—graduated income tax, centralization of much economic control in the hands of the State, nationalization of several major industries including transport, free education for all children in state schools. The unrestrained capitalist system as Marx knew it in the nineteenth century has everywhere ceased to exist, and this has happened by step-by-step reform, not by once-for-all revolution. This is not to say that the existing system is perfect—far from it. But it is to suggest that Marx's rejection of any idea of gradual reform is radically mistaken; and reflection on the suffering and violence involved in real revolutions may confirm this.

Like Christianity, Marx envisages a total regeneration of man, but he expects it entirely within this secular world. Communism is 'the solution to the riddle of history' (p. 250), for the abolition of private property is supposed to ensure the disappearance of alienation and the coming of a genuinely classless society. Marx is very vague on how all this will be achieved, but he suggests that there will be an intermediate period during which the transition will take place, and that this will require 'the dictatorship of the proletariat' for its accomplishment (p. 261). But in the higher phase of communist society, the State will wither away, and the true realm of freedom will begin. Then human potentiality can develop for its own sake (p. 260), and the guiding principle can be: 'From each according to his ability, to each according to his needs' (p. 263).

Some of this utopian vision must surely be judged wildly unrealistic. Marx gives us no good reason to believe that communist society will be genuinely classless, that those who exercise the dictatorship of the proletariat will not form a new governing class with many opportunities to abuse their power, as the history of Russia since the revolution obviously suggests. There is no ground for expecting any set of economic changes to eliminate *all* conflicts of interest forever. The State, far from withering away, has become ever more powerful in communist countries (perhaps the very nature of modern industry and technology makes this inevitable).

Yet with other elements in Marx's vision, we cannot but agree. The idea of a decentralized society in which men co-operate in communities for the common good, the application of science and technology to produce enough for all, the shortening of the working day so that men can increasingly choose to spend their leisure time in the free development of their potential, the idea of a society in balance with nature—all these are ideals which almost everyone will share, even though it is not clear that they are compatible. No doubt, it is because Marxism offers this kind of hopeful vision of the future that it can still win and retain the allegiance of so many people. For despite the obvious defects of life in the existing communist countries, many of their inhabitants maintain a genuine belief in Marxist theory. And despite the reforms that have already altered the face of capitalism, many people in the West see the need for a further transformation of the existing socioeconomic system and look to Marx for inspiration for such a change.

Like Christianity, Marxism is more than a theory, and the disputability of many of its theoretical assertions does not make it lie down and die. It contains a recipe for social salvation and offers a critique of any existing society. However, Marx's emphasis on social and economic factors directs our attention to one, but only one, of the obstacles in the way of human progress. We must look elsewhere, for instance to Freud, for more about the nature and problems of human *individuals*.

For Further Reading

Basic text: *Karl Marx: Selected Writings in Sociology and Social Philosophy*, translated by T. B. Bottomore, edited by T. B. Bottomore and M. Rubel (Penguin, London, 1963; McGraw-Hill paperback, New York, 1964). Perhaps the best selection of readings from all periods of Marx's work. It has a bibliography of Marx's principal works at the end.

Exploration of Marx's writings could continue with the volumes in the Pelican Marx library, which includes the first complete translation of the *Grundrisse*. David McLellan's *The Thought of Karl Marx* (Macmillan, London, 1971; Harper & Row, New York, 1972) is a useful guide.

For one of the most famous of the many critical studies, see Karl Popper's *The Open Society and Its Enemies,* Volume II (Routledge & Kegan Paul, London, 5th edn. 1966; Princeton University Press paperback, Princeton, N.J., 1966). Readers will recognize in it the source of many of my points.

For a readable biography of Marx, placing most emphasis on the development of his ideas, see Sir Isaiah Berlin's *Karl Marx: His Life and Environment* (Oxford University Press, London, 3rd edn. 1963; Oxford University Press Galaxy Books paperback, New York, 1978). This also contains further bibliography.

For a deeper discussion of Marx on human nature, see J. Plamenatz, *Karl Marx's Theory of Man* (Oxford University Press, New York, 1975). For an encyclopedic critical guide to Marx and the many subsequent varieties of Marxism, see L. Kolakowski, *Main Currents of Marxism* (Oxford University Press, New York, 1975, in three volumes). For a sophisticated defence of Marxism in light of modern analytical philosophy, see G. A. Cohen, *Karl Marx's Theory of History: A Defence* (Oxford University Press, New York, 1978).

Page references in this chapter are to the Penguin edition of *Karl Marx: Selected Writings in Sociology and Social Philosophy*. Readers of the McGraw-Hill paperback edition should use the references below.

Stevenson	*McGraw-Hill paperback*
page 58 refers to pages	70; 52; 141–2
59	51; 54, 75, 98; 51
60	64–5, 138–43, 186, 201, 231–3
61	67; 53, 70–1; 68
62	76–7, 245; 61–2; 77; 51; 53; 169; 247
63	169; 244; 168
64	169–70; 171–2; 97; 217; 220
65	243
66	244; 256; 255; 258

6

Freud: Psychoanalysis

The next theory of human nature I should like to consider is that of Freud, for it is commonplace (but still true) to say that he has revolutionized our understanding of ourselves in this century. So no adequate discussion of human nature can fail to grapple with his thought. But this is a peculiarly difficult task to attempt in a single chapter, even if we concentrate only on Freud himself and ignore the many later developments in psychoanalytic theory and practice. For Freud spent nearly fifty years developing and modifying his theories, writing so vast an amount of material that the non-specialist cannot hope to read it all. What I shall attempt to do here is to sketch briefly his life and work (for this will suggest some of the complexity and richness of the latter), then summarize some of the most fundamental points of his theory, diagnosis, and prescription, and finally raise some of the main critical questions which must be asked.

Life and Work

Sigmund Freud was born in Moravia in 1856, but in 1860 his family moved to Vienna, where he lived and worked until the last year of his life. Even in his schooldays his interests extended to the whole of human life, and when he entered the University of Vienna as a medical student he did not immediately concentrate on medicine. He became interested in biology, however, and spent six years doing research in physiology in the laboratory of the great German scientist Brücke. In 1882 he became engaged to be married, and therefore needed a career with a more secure financial reward, so he unwillingly began work as a doctor in the Vienna General Hospital. In 1886 he was able to marry, and to set up the private practice in 'nervous diseases' which he maintained until the end of his life.

Freud's working life from then on can be roughly divided into three phases. In the first of these he made his great discoveries and developed the theory and treatment which has come to be known as psychoanalysis. His interest in psychological problems had been fired by a visit to Paris in 1885–6 to study under Charcot, a French neurologist who was then using hypnotism to treat hysteria. Faced with similar symptoms in his own patients, Freud experimented with both electrotherapy and hypnotic suggestion, but found them unsatisfactory, so he began to try another method which had been used by Breuer, a senior Viennese consultant who was a friend of his. Breuer's approach was based on the assumption that hysteria was caused by some intense emotional experience that the patient had forgotten; the treatment was to induce the recall of the experience, and hence a discharge of the corresponding emotion. This idea that people could suffer from some idea or memory or emotion of which they were not conscious, but from which they could be cured by somehow bringing it into consciousness, is the basis from which Freud's psychoanalysis developed. He went on to introduce the concepts of resistance, repression, and transference, which we shall look at soon. In 1895 he published *Studies on Hysteria* jointly with Breuer, but soon afterwards he broke with his friend and went on his own theoretical way. At the same time he was writing (to Fliess, another friend who influenced him in this period) the *Project for a Scientific Psychology,* a manuscript which was not published until 1950. In this he tried, somewhat abortively, to relate the psychological theory he was then developing to a material basis in the physiology of the brain, which he had already studied in his premedical work. In the last few years of the century he undertook the difficult task of analysing his own mind, and arrived at the concepts of infantile sexuality and of interpreting dreams, both of which are central to the mature theory of psychoanalysis.

The second phase of Freud's work, in which the great books expounding this mature theory appeared, can conveniently be dated from the publication in 1900 of *The Interpretation of Dreams,* the work which he himself regarded as his best. There followed in 1901 *The Psychopathology of Everyday Life,* in which he analysed the unconscious causation of everyday errors, and in 1905 *Three Essays on the Theory of Sexuality.* These three works applied psychoanalytic theory to the whole of normal mental life, not just to pathological conditions. The international recognition and spread of psychoanalysis began, and in 1909 Freud was invited to America where he delivered the lectures which formed the first of his expository works, the *Five Lectures on Psycho-Analysis.* In 1913–14 came *Totem and Taboo,* applying his theories to anthropological material, and in 1915–17 he gave the *Introductory Lectures on Psycho-Analysis* at the University of Vienna, in which he expounded at length the complete theory as it had developed up till then.

From after the end of the First World War until his death, the third phase of Freud's work included further developments and changes in theory, together with wide-ranging speculative attempts to apply his theories to social questions. During this period he suffered more and more from the cancer which eventually killed him. In 1920 came *Beyond the Pleasure Principle,* in which he first introduced the concept of a death instinct independent of the erotic or life instinct which he had long before postulated. Another late development was the tripartite structure of the mind–id, ego, and superego–given in 1923 in *The Ego and the Id.* Most of Freud's last years were devoted to sociological works. *The Future of an Illusion* of 1927 is about religion, treating it, as the title suggests, as false beliefs whose origin must be explained psychologically. *Civilization and Its Discontents* (1930) discusses the conflict between the demands of civilized society and the instincts implanted in every person, and *Moses and Monotheism* (1939) discusses the history of the Jews from a psychoanalytic point of view. In 1938 Hitler invaded Austria, but Freud was allowed to leave for London, where he spent the last year of his life writing a brief exposition of the latest version of his doctrine—*An Outline of Psycho-Analysis.*

There is no better introduction to Freud's thought than his own *Five Lectures* of 1909. Accordingly, in what follows I shall make my page references to the Pelican book *Two Short Accounts of Psycho-Analysis,* which contains these lectures together with another expository work, *The Question of Lay Analysis* of 1926. In the latter, Freud discussed whether non-doctors should be allowed to practise psychoanalysis, and explained his basic theory in terms of the tripartite structure which he had recently formulated. (For American references see the end of the chapter.)

Theory of the Universe

What is distinctive in Freud's thought comes not under this heading but in his theory of man, but we should take note of his background assumptions and how they differ from those of the theories discussed so far. Freud started his career as a physiologist and claimed to remain a scientist throughout and to treat all phenomena, including those of human nature, scientifically (p. 100). He made no theological assumptions (he was in fact a convinced atheist), nor any metaphysical assumptions like Plato on the Forms, or Marx on the movement of history. What he did assume (no doubt from his training in nineteenth-century science and his research in Brücke's Physiological Laboratory) is that all phenomena are determined by the laws of physics and chemistry, and that even man himself is a product of natural evolution, ultimately subject to the same laws.

Theory of Man

I shall try to summarize Freud's basic concepts under four main heads. The first is a strict application of the principle of determinism–that every event has preceding sufficient causes–within the realm of the mental. Things that had formerly been assumed to be of no significance for understanding a person, such as slips of the tongue, faulty actions, and dreams, Freud took as determined by hidden causes in a person's mind. So they could be highly significant, revealing in disguised form what would otherwise remain unknown (pp. 56, 60, 65–6). Nothing which a person does or says is really haphazard or accidental; everything can in principle be traced to causes which are somehow in the person's mind. This might seem to imply a denial of human free will, for even when we think we are choosing perfectly freely and even arbitrarily, Freud could say that there are uncontrollable causes determining our choice. There is an interesting parallel with Marx here, for both are saying that our consciousness, far from being perfectly 'free' and 'rational,' is really determined by causes of which we are not aware; but whereas Marx says that these causes are social and economic in nature, Freud claims that they are individual and mental.

The second main point–the postulation of *unconscious* mental states–thus arises out of the first. But we must be careful to understand this concept of the unconscious correctly. There are lots of mental entities, for instance memories of particular experiences or of particular facts, of which we are (fortunately!) not continually conscious, but which we can call to mind whenever necessary. These Freud would call 'preconscious' (that which can become conscious), reserving the term 'unconscious' for that which *cannot* become conscious at all under normal circumstances. The assertion is, then, that the mind is not co-extensive with what is conscious or can become conscious, but includes items of which the person can have no ordinary knowledge at all (pp. 107, 43, 47). To use a familiar but helpful analogy, the mind is like an iceberg, with only a small proportion of it visible above the surface, but a vast hidden bulk exerting its influence on the rest. For the unconscious is *dynamic* in nature, that is, it actively exerts pressures and influences on what a person is and does. For instance, there are unconscious desires, which can cause someone to do things that he cannot explain rationally, to others or even to himself.

In asserting the existence of unconscious mental entities, is Freud committed to a dualism between mind and body, or between mental states and physical states? I think this would be a misinterpretation of him. Many philosophers would now agree that in talking of ordinary *conscious* mental states (e.g., thoughts, wishes, emotions) we are not thereby committed to a metaphysical dualism. The question of the ultimate nature of such mental states is a philosophical problem which is left open by our everyday language about them. There is no reason to suppose that the case is any different for unconscious mental states. Freud himself, with his physiological training, would certainly have rejected any suggestion of dualism. After his early attempt to find a physiological basis for his psychological theories (in the *Project* of 1895), he came to the view that such matters are not of psychological interest (p. 103) and was content to leave them to the future development of neurophysiology. But that all the mental entities he postulated had *some* physiological basis, although as yet unknown, he did not doubt. So his theory of man is not a dualist one like Plato's.

There is an interesting parallel with Plato, however, in the theory of the tripartite structure of the mind, which Freud introduced in his later expositions in the 1920s. This is not the same as the distinction between conscious, preconscious, and unconscious which he had used until then. He now distinguished three major structural systems within the human mind or personality: the *id,* which contains all the instinctual drives seeking immediate satisfaction; the *ego,* which deals with the real world outside the person, mediating between it and the id (pp. 104–5); and the *superego,* a special part of the ego which contains the conscience, the social norms acquired in childhood (p. 137). The superego has a connection with the id also, for it can confront the ego with moral rules like a strict father; the ego has to reconcile the conflicting demands of id, superego, and external reality. Whatever can become conscious is in the ego, although even in it there may be things which remain unconscious, whereas everything in the id is permanently unconscious (p. 108). So the id corresponds closely to Plato's element of Appetite or desire, but it is not so clear how ego and superego correspond to plato's elements of Reason and Spirit. In its reality-knowing function the ego would seem to be akin to the Reason, but Reason for Plato has also a moral function which Freud would give to the super-ego. And yet the spirited element seems to be performing a moralistic function in the situation of self-disgust by which Plato illustrated it.

The third main feature of Freud's concept of man is his theory of the instincts or 'drives'—or rather, his several theories of them, for this is one of the most variable parts of his work. The instincts are the motive forces within the mental apparatus, all the 'energy' in our minds comes from them alone (p. 110). (Freud used this mechanical or electrical language in an almost literal way, influenced no doubt by his scientific training and his *Project* of 1895.) Although he held that one can distinguish an indeterminate number of instincts, he also thought that they could all be derived from a few basic ones, which can combine or even replace each other in multifarious ways. Now he undoubtedly held that one such basic instinct is sexual in nature (p. 69); but it is a vulgar misinterpretation of Freud to say that he traced *all* human behaviour to sexual motivations. What *is* true is that he gave sexuality a much wider scope in human life than had formerly been recognized (p. 76). He claimed that sexual instincts exist in children from birth onwards (pp. 71, 121), and he asserted the crucial importance of sexual energy or 'libido' in adult life (p. 118). But Freud always held that there was at least one other basic instinct or group of instincts. In his early work he talked of the self-preservative instincts such as hunger, and contrasted these with the erotic instincts, one unusual aggressive manifestation of which was sadism. But in his later work from about 1920 onwards he changed the classification, putting erotic and self-preservative instincts into one basic 'Life' instinct (Eros), and referring sadism, aggression, self-destruction, etc., to a basic 'Death' instinct (Thanatos).

The fourth main point is Freud's developmental or historical theory of individual human character. This is not just the rather obvious truism that personality depends on experience as well as on hereditary endowment. Freud started from Breuer's discovery that particular 'traumatic' experiences could, although apparently forgotten, continue to exercise a baneful influence on a person's mental health (p. 39). The fully fledged theory of psychoanalysis generalizes from this and asserts the crucial importance, for adult character, of the experiences of infancy and early childhood (pp. 70, 115). The first five years or so are held to be the time in which the basis of individual personality is laid down. So one cannot fully understand a person until one comes to know the psychologically crucial facts about his early childhood.

Freud produced detailed theories of the stages of development through which every child grows (pp. 73, 121). These particular theories can of course be distinguished from the general developmental approach, and are more easily tested by observation. They are specifically concerned with the development of sexuality; Freud widened the concept of sexuality to include any kind of pleasure obtained from parts of the body. He suggested that infants first obtain such pleasure from the mouth (the oral stage) and then from the other end of the alimentary tract (the anal stage). Both boys and girls then become interested in the male sexual organ (the phallic stage). The little boy is alleged to feel sexual desires for his mother, and to fear castration by his father (the situation of the 'Oedipus complex'; p. 125). Both desire for mother and hostility to father are then normally repressed. Freud supposed that the little girl develops 'penis envy' at the same stage; but he rarely treated feminine sexuality so thoroughly. From age five until puberty (the 'latency' period) sexuality is much less apparent. It returns in its full 'genital' development at the beginning of adulthood.

Diagnosis

Like Plato, Freud says that individual well-being or mental health depends on a harmonious relationship between the various parts of the mind, and between the person and the real world in which he has to live. The ego has to reconcile id, superego, and external world, perceiving and choosing opportunities for satisfying the instinctual demands of the id without transgressing the standards required by the superego (pp. 111, 137). If the world is unsuitable and does not give any such opportunities, then of course suffering will result, but even when the environment is reasonably favourable, there will be mental disturbance if there is inner conflict between the parts of the mind. So neurosis results from the frustration of basic instincts, either because of external obstacles or because of internal mental imbalance (p. 80).

There is one particular mental misadaptation which is of crucial importance in the causation of neurotic illnesses, and this is what Freud called repression. In a situation of extreme mental conflict, where a person experiences an instinctual impulse which is sharply incompatible with the standards he feels he must adhere to, it is possible for him to put it out of consciousness (pp. 48–9), to flee from it, to pretend that it does not exist (p. 113). So repression is one of the so-called 'defence mechanisms,' by which a person attempts to avoid inner conflicts. But it is essentially an escape, a pretence, a withdrawal from reality (p. 80), and as such is doomed to failure. For what is repressed does not really disappear, but continues to exist in the unconscious portion of the mind. It retains all its instinctual energy, and exerts its influence by sending into consciousness a disguised substitute for itself—a neurotic symptom (pp. 52, 114). Thus the person can find himself behaving in ways which he will admit are irrational, yet which he feels compelled to continue without knowing why. For by repressing something out of his consciousness he has given up effective control over it; he can neither get rid of the symptoms it is causing, nor voluntarily lift the repression and recall it to consciousness.

As we should expect from his developmental approach to the individual, Freud locates the decisive repressions in early childhood (p. 115). And as we might expect from his emphasis on sexuality, he holds them to be basically sexual (p. 71). It is essential for the future mental health of the adult that the child successfully passes through the normal stages of development of sexuality. But this does not always proceed smoothly, and any hitch in it leaves a predisposition to future neurosis (pp. 75, 122); the various forms of sexual perversion can be traced to such a cause. One typical kind of neurosis consists in what Freud called 'regression' (p. 75), the return to one of the stages at which childish satisfaction was obtained. He even identified certain adult character-types as oral and anal, by reference to the childhood stages from which he thought they originated.

There is much more detail in Freud's theories of the neuroses, into which we cannot enter here, but we have already noted that he can attribute part of the blame for them on the external world, and so we should look a bit more at this social aspect of his diagnosis. For the standards to which a person feels he must conform are one of the crucial factors in mental conflict, but these standards are (in Freud's view) a product of the person's social environment–primarily his parents, but including anyone who has exerted influence and authority on the growing child. It is the instillation of such standards that constitutes the essence of education, and makes a child into a member of civilized society; for civilization requires a certain control of the instincts, a sacrifice of instinctual satisfaction (p. 81) in order to make cultural achievements possible (p. 86). But the standards instilled are not automatically the 'best' or most rational or most conducive to individual happiness. Certainly, individual parents vary widely, and maladjusted parents will be likely to produce maladjusted children. But Freud was prepared to entertain the possibility (most obviously in his late work *Civilization and Its Discontents,* although the beginnings of this line of thought are apparent in much earlier works) that the whole relationship between society and the individual has got out of balance, that our whole civilized life might be neurotic (p. 119). Even as early as the *Five Lectures* of 1909, he asserted that our civilized standards make life too difficult for most people and that we cannot deny a certain amount of satisfaction to our instinctual impulses (pp. 86–7). So there is a basis in the writings of

Freud himself for those later Freudians who diagnose the main trouble as lying in society rather than in the individual.

Prescription

As usual, prescription follows from diagnosis. Freud's aim was to restore a harmonious balance between the parts of the mind, and between the individual and his world. The latter might well involve programmes of social reform, but Freud never specified these in any detail; his everyday practice was the treatment of neurotic patients by psychoanalysis. The word 'psychoanalysis' refers at least as much to Freud's method of treatment as to the theories on which that treatment is based. It is this therapy which we must now examine.

The method developed gradually out of Breuer's initial discovery that one particular hysterical patient could be helped by being encouraged to talk about the fantasies which had been filling her mind, and could actually be cured if she could be induced to remember the 'traumatic' experiences that had apparently caused her illness in the first place (pp. 33–9). Freud started using this 'talking cure,' and, assuming that the pathogenic memories were always somewhere in the person's mind even if not ordinarily available to consciousness, he asked his patients to talk freely and uninhibitedly, hoping that he could interpret the unconscious forces behind what they said (pp. 58–9). He encouraged them to say *whatever* came into their mind, however absurd (the method of 'free association'). But he often found that the flow of associations would dry up, and the patient would claim to know nothing more, and might even object to further inquiry. When such 'resistance' happened, Freud took it as a sign that he was really getting near the correct interpretation of the repressed complex. He thought that the patient's unconscious mind would somehow realize this and try to prevent the painful truth being brought into consciousness (p. 48). Yet only if the repressed material could be brought back into consciousness could the patient be cured, and his ego given back the power over the id which it had lost in the process of repression (pp. 66, 115).

But to achieve this happy result could take a long process, involving perhaps weekly sessions over a period of years. The analyst must try to arrive at the correct interpretations of his patient's condition, and present them at such a time and in such a way that the patient can accept them (p. 134). The patient's dreams will provide very fruitful material for interpretation, for according to Freud's theory the 'manifest' content of a dream is always the *disguised* fulfillment of repressed wishes, which are its real or 'latent' content (pp. 60–4). Faulty actions can also be interpreted to reveal their unconscious causation (p. 65). As one would expect from the theory we have summarized, the interpretations will very often refer to a person's sexual life, his childhood experiences, his infantile sexuality, and his relationships to his parents. Clearly all this demands a relationship of peculiar confidence between patient and analyst, but Freud found that much more than this happened; in fact, his patients manifested a degree of emotion toward him that could almost be called falling in love. This phenomenon he labelled 'transference,' on the assumption that the emotion was somehow transferred to the analyst from the real-life situations in which it was once present, or from the unconscious fantasies of the patient (pp. 82–3, 139–41). The handling of such transference is of crucial importance for the success of the analysis, for it itself can be analysed and traced back to its sources in the patient's unconscious (pp. 141–2).

The goal of psychoanalytic treatment can be summarized as self-knowledge. What the cured neurotic does with his new self-understanding is up to him, and various outcomes are possible. He may replace the unhealthy repression of instincts by a rational, conscious, control of them (*sup*pression rather than *re*pression); or he may be able to divert them into acceptable channels (sublimation); or he may decide that they should be satisfied after all (pp. 185–6). But there is no possibility at all of a result that is sometimes feared by the layman–that primitive instincts when unleashed will take over completely–for their power is actually *reduced* by being brought into consciousness (pp. 84–5).

Freud spent his life treating individual neurotic patients. But he never thought that psychoanalytic treatment is the answer to *every* human problem. When grappling speculatively with the problems of civilization and society he was realistic enough to realize their extreme complexity and to abstain from offering any panacea. But he did hold that psychoanalysis had much wider applications than just the treatment of neurotics (p. 168). He said 'our civilization imposes an almost intolerable pressure on us and it calls for a corrective,' and speculated that psychoanalysis might help to prepare such a corrective (pp. 169–70). At the end of *Civilization and Its Discontents* he cautiously proposed an analogy between cultures and individuals, so that cultures too might be 'neurotic.' But he recognized the precariousness of the analogy, and refused to 'rise up before his fellow-men as a prophet.'

Critical Discussion

The position of psychoanalysis on the intellectual map is still a matter of dispute. Psychoanalysts continue to practise, with a variety of Freudian and post-Freudian doctrines. But many academic psychologists and some practising psychiatrists condemn psychoanalysis as almost totally unscientific, as more akin to witchcraft than to respectable scientific medicine. Some critics fasten on the cult-like orthodoxy imposed by the various schools of psychoanalysts, and the 'indoctrination' which every aspirant psychoanalyst is required to go through (by being analysed himself). They therefore classify the theory and practice as that of a quasi-religious faith. It certainly has a method of disparagingly analysing the motivations of its critics (for any questioning of the truth of psychoanalytic theory can be alleged to be based on the unconscious resistance of the critic to its unpleasant implications). So if (as many say) the theory also has a built-in method of explaining away any evidence which appears to falsify it, then it will indeed be a closed system, in the sense defined in Chapter 2. And since belief in the theory is a requirement of membership of psychoanalytic institutes, it might even be said to be the ideology of those social groups. However, we should look more closely at the case before we pass judgement.

We must first distinguish two independent questions: the truth of Freud's theories, and the effectiveness of the method of treatment based on them. On the status of the theories, the crucial problem is whether they are falsifiable, for we have seen that Freud claimed his theories to be scientific ones, and we have taken empirical falsifiability as a necessary condition for scientific status (in Chapter 2). But for some of the central propositions of Freudian theory it is not clear whether they are falsifiable at all. Let us illustrate this with examples from the four main sections in which we summarized the theory. The postulate of psychic determinism leads to specific propositions such as that all dreams are disguised wish-fulfillments. Can this be tested? Where an interpretation in terms of an independently established wish of the dreamer is proposed and accepted, well and good. But what if no such interpretation is found? A convinced Freudian can still maintain that there is a wish whose disguise has not been seen through. But this would make it impossible to show that a dream is *not* the disguised fulfilment of a wish, and would therefore evacuate the proposition of any genuine empirical content, leaving only the prescription that we should *look* for a wish fulfilled by the dream. The proposition can be empirical only if we can have independent evidence for the existence of the wish and the correct interpretation of its disguise.

Consider next the postulation of unconscious mental states, and in particular the tripartite structure of id, ego, and superego. Freud did not of course expect these entities to be visible or tangible, and we have seen that he dismissed the question of what they are made of as of no psychological interest (p. 103). But that was because he had then abandoned any hope of discovering a physiological basis in the brain for such mental factors. It remains logically possible that neurophysiology may progress to a point where we could identify three physical systems in the brain which play the roles ascribed by Freud to id, ego, and superego. But as yet there is no sign of such a possibility being realized. In the meantime we must ask ourselves whether the postulation of unconscious mental states offers any genuine explanation of human behaviour. It would be too quick a move to dismiss them just because they are unobservable, for scientific theory often postulates entities which are not directly evident to any of the human senses–for example, atoms, electrons, magnetic fields, and radio waves. But in these cases there are clear 'correspondence rules' connecting the unobservable entities with observable phenomena, so that, for instance, we can infer the presence or absence of a magnetic field from the visible behaviour of a compass needle. The trouble with many of the Freudian entities is that there are no such unambiguous rules for them. Stamp collecting may be asserted to be a sign of unconscious 'anal retentiveness,' but could one show that such an unconscious trait is *not* there in someone?

Freud's theory of instincts is perhaps the part which is least open to empirical testing, as is suggested by his vacillations on the subject. One can describe as instinctive any form of behaviour that is not learned from the experience of the individual (although it may often be difficult to *show* that it has not been learned). But nothing seems to be added by referring instinctive behaviour to *an* instinct as its cause–for what evidence can there be for the existence of an instinct except the occurrence of the corresponding unlearned behavior? And if it is claimed that there are only a certain number of basic instincts, it is not at all clear how one could decide which are basic, and how they are to be distinguished and counted. Could any evidence of human behaviour settle whether either of Freud's main instinct theories is right, as against, say, an Adlerian theory of a basic instinct of self-assertion or a Jungian theory of an instinctual need for God? Once again, the postulation of unobservables is useless unless testable by observation.

The developmental approach to individual character and the theory of the stages of infantile sexual development are more easily tested by observation. In this area, some of Freud's propositions are clearly supported by the evidence, others are neither supported nor refuted, while others are very difficult to test (see Kline's book recommended below). The *existence* of what Freud called the oral and anal characters has been confirmed by the discovery that certain traits of character (for instance, parsimony, orderliness, and obstinacy—the anal traits) do tend to go together. But the theory that these types of character *arise* from certain kinds of infant-rearing procedure has not been supported by the available studies. However, there are practical difficulties in getting the necessary correlations between infantile experience and adult character, so the theory is not yet refuted. For some other parts of Freud's psychosexual theories there are conceptual difficulties about testing. How, for example, could one test whether infants get erotic pleasure from sucking? Some studies suggest that infants who have less opportunity to suck during feeding tend to do more sucking of their thumbs—but can this really be construed as evidence for the *erotic* nature of sucking?

This rather brief treatment of a few examples does at least show that there is serious doubt about the scientific status of some of Freud's key theoretical assertions. Some seem unfalsifiable, while of those than *can* be tested only some have received clear support from the evidence. The testing is subject to a mixture of complex practical and conceptual difficulties.

It has been suggested (by both psychoanalysts and philosophers) that psychoanalysis is not primarily a set of assertions to be tested empirically, but more a way of understanding people, of seeing a *meaning* in their actions, their mistakes, their jokes, and their dreams. It may be said that, since human beings are vastly different from the entities studied by physics and chemistry, one should not condemn psychoanalysis for failing to meet criteria for scientific status which have been taken from the established physical sciences. Perhaps the psychoanalytic discussion of a dream is more akin to literary criticism, such as the interpretation of an obscure poem, in which there are reasons (but not conclusive reasons) for a variety of interpretations. Many of Freud's conceptions can be seen as extensions of our ordinary ways of understanding each other in terms of everyday concepts such as love, hate, fear, anxiety, rivalry, etc. And the experienced psychoanalyst can be seen as someone who has acquired a deep understanding of the springs of human motivation and a skill in interpreting the multifarious complexities of how they work out in particular situations, regardless of the theoretical views he may espouse.

Such a view of psychoanalysis has been given philosophical backing by a sharp distinction between *reasons* and *causes,* and hence between scientific explanation (in terms of causes) and the explanation of human actions (in terms of reasons, i.e., the beliefs and desires which made it rational for the agent to do what he did). And it has been suggested that Freud misunderstood the nature of his own theories, in so far as he presented them as offering new scientific discoveries about the causes of human behaviour. However, this dichotomy has come in for criticism from those who argue that someone's conscious beliefs and desires can indeed be causes of his actions, and hence that *unconscious* beliefs and desires may well play this causal role too. There are deep philosophical issues here about how far the methods of scientific investigation and explanation are applicable to human beliefs and actions, but it is impossible to pursue them here (for some further reading see notes 10 and 11 to Chapter 10).

Doubts about psychoanalytic theory naturally extend to the treatment based on it. But when that treatment has already been widely applied, we should be able to form some estimate of its effectiveness. This would in principle give a further test of the theory, for if it is really true we would expect the treatment to be effective. However, matters are not easy here, either. For one thing, a true theory might be badly applied in practice; and for another, there is doubt about what constitutes 'cure' from neurotic illness. A proportion of two-thirds has been given as the approximate rate of cure for patients who complete psychoanalytic therapy. This may seem favourable, but of course it must be compared with a 'control group' of similar cases who have been treated by other psychiatric methods or not treated at all. The proportion of recovery in such groups is also said to be about two-thirds, so there is no proof of any therapeutic effectiveness.

Thus no clear verdict can be passed on Freud's theories as a whole. His genius is undisputed, but however influential a man's thought, we can never be excused the task of critically evaluating what he says. Freud said so much of such great importance that the digesting and testing of it will occupy both philosophers and psychologists for many years to come.

For Further Reading

Basic text: *Two Short Accounts of Psycho-Analysis,* translated and edited by James Strachey (Penguin, London, 1962). This book is not available in the United States, but most of the first 'short account,' "Five Lectures on Psycho-Analyses," is reprinted in *A General Selection from the Works of Sigmund Freud,* edited by John Rickman (Doubleday Anchor paperback, New York, 1957), and much of the second 'short account,' "The Question of Lay-Analysis," is reprinted in *The Study of Human Nature,* edited by Leslie Stevenson (Oxford University Press, New York, 1981). Page references to these anthologies (labelled 'R' and 'S') are supplied below.

Exploration of Freud's own writings could continue with the *Introductory Lectures on Psycho-Analysis* of 1915–17, and then with the other volumes in the Pelican Freud library.

For a clear general evaluation of Freud's achievement, see *The Standing of Psycho-Analysis,* by B. A. Farrell (Oxford University Press, New York, 1981); and for details of the empirical testing of Freudian theories, see *Fact and Fantasy in Freudian Theory,* by Paul Kline (Methuen, London, 2nd edn. 1981).

For biography, see the famous *Life and Work of Sigmund Freud,* by Ernest Jones, conveniently edited and abridged in one volume by Lionel Trilling and Steven Marcus (Penguin, London, 1964; Basic Books, New York, 1961); also more recently *Freud, Biologist of the Mind,* by Frank J. Sulloway (Basic Books, New York, 1979; Fontana, London, 1980).

For an introduction to the later developments in psychoanalytic theory, see *Freud and the Post-Freudians,* by J. A. C. Brown (Penguin, London and New York, 1964).

For recent discussion of the philosophical issues involved, see *Philosophical Essays on Freud,* edited by R. Wollheim and J. Hopkins (Cambridge University Press, New York, 1982).

This book	*Anthologies by Rickman and Stevenson*
page 72 refers to page	S 16
73	R 18, 21–3, 25–6, S 170, R 10, 12
74	S 167, 168, 186, 170
75	S 171, R 27, 32, 29, S 176
76	R 16, 28, S 175, R 29–31, S 177–80
77	S 171–2, 186, R 35–6, 13, S 173–4, R 36, 16, S 174, 175, R 28
78	31, S 177, R 37, 42, S 192
79	R 42–3, 5–7, 20–1, 13, S 175
80	183, R 21–5, 25–6, R 38–9, S 188–90, R 40–42
81	S 191–2
82	S 167

7

Sartre: Atheistic Existentialism

In moving from Freud to Sartre we go from the Vienna of the turn of the century to the Paris of the 1930s and 1940s, from the psychological side of medicine to a philosophy expressed in imaginative literature as well as in academic exposition. Yet there is a common concern with the problems of the human individual, and in particular with the nature of consciousness. We shall find that Sartre too has a fourfold theory of human nature, but we should first put him in the context of existentialism as a whole.

Many writers, philosophers, and theologians have been called 'existentialist.' In so far as a common core can be discerned there would seem to be three main concerns which are central to existentialism. The first is with the *individual* human being, rather than with general theories about him. Such theories, it is thought, leave out what is most important about each individual–his uniqueness. Secondly, there is a concern with the *meaning* or purpose of human lives, rather than with scientific or metaphysical truths about the universe. So inner or subjective experience is typically regarded as more important than 'objective' truth. Thirdly, the concern is with the *freedom* of individuals as their most important and distinctively human property. So existentialists believe in the ability of every person to choose for himself his attitudes, purposes, values, and way of life. And they are concerned not just to maintain this as a truth, but to persuade everyone to act on it. For in their view the only 'authentic' and genuine way of life is that freely chosen by each individual for himself. These three concerns, then, are really aspects of one basic theme.

But this common core of existentialism can be found in a wide variety of contexts. It is naturally expressed in descriptions of the concrete detail of particular characters and situations, as in plays and novels. However someone can count as an existentialist *philosopher* only if he makes some attempt at *general* statement about the human condition (even if that statement consists in denying the possibility or the importance of other general statements!). Existentialist philosophies come in various forms, the most radical division being between the religious and the atheistic. The Danish Christian thinker Kierkegaard (1813–55) is generally regarded as the first modern existentialist. Like Marx, he reacted against the Hegelian system of philosophy, but in a quite different direction. He rejected the abstract theoretical system as like a vast mansion in which one does not actually live, and maintained instead the supreme importance of the individual and his choices. Distinguishing three main ways of life–the aesthetic, the ethical, and the religious–he required each individual to choose between them. But he also held that the religious way (more specifically, the Christian one) is the highest, although it can be reached only by a free 'leap into the arms of God.' The other great nineteenth-century source of existentialism, the German writer Nietzsche (1844–1900), was aggressively atheist. He held that since 'God is dead' (i.e., the illusions of religion have been seen through) we will have to rethink the whole foundations of our lives, and find their meaning and purpose in human terms alone. In this, he had much in common with his earlier compatriot Feuerbach, whose humanistic atheism I mentioned briefly when introducing Marx. What is more distinctive of Nietzsche is his emphasis on our freedom to change the basis of our values, and his vision of the 'Super-man' of the future, who will reject our present meek religiously based values by more real ones based on the human 'will to power.'

In the twentieth century, too, existentialists have included both Christians and atheists. Existentialism has been a major force in theology, both Protestant and Catholic, as well as in philosophy. The philosophical movement has been centred on the mainland of Europe, especially in Germany and France, and has had much less influence in English-speaking countries. Its sources can be traced to Kierkegaard and Nietzsche, but also to the 'phenomenology' of the German-speaking philosopher Husserl (1859–1938). This somewhat obscure philosophical method tried to find an unproblematical starting-point by describing only the 'phenomena' as they seem to be, without making any assumptions about what they really are. So it gave philosophy a subjective, quasi-psychological twist, making it the study of human consciousness. It is this concern with consciousness that we find in twentieth-century existentialist philosophers. In Germany the most important of these is Heidegger (born in 1889), whose major work, the huge and obscure *Being and Time,* appeared in 1927. His central concern is however with human existence, and the possibility of 'authentic' life through facing up to one's real position in the world, and in particular to the inevitability of one's death.

But I do not wish to deal here with existentialism in general; the above has been sketched as background for our consideration of the most famous of the French existentialists, Jean-Paul Sartre (1905–1980). In a brilliant educational career he absorbed, among much else, the thought of the great European philosophers, especially that of Hegel, Husserl, and Heidegger. Many of the obscurities of Sartre's own philosophical style can be traced to the influence of these three writers of ponderous German abstractions. Themes from Husserl's work can be detected in his first books—the novel *Nausea* of 1938, and three philosophical studies of psychological topics: *Imagination* (1936), *Sketch for a Theory of the Emotions* (1939), and *The Psychology of the Imagination* (1940). His central work, expounding at length his philosophy of human existence, is the celebrated *Being and Nothingness,* first published in 1943. He gave a much shorter and clearer account of atheistic existentialism in *Existentialism and Humanism,* a lecture delivered in Paris in 1945, but unfortunately his treatment there is popular and superficial, and should not be relied upon as an exposition of his thought. During the Second World War he was active in the French Resistance to the Nazi occupation, and some of the atmosphere of that time can be found in his work; for instance the choice which confronted all Frenchmen of collaboration, resistance, or quiet self-preservation was a very obvious particular case of what existentialists see as the ever-present necessity for individual choice. Such themes are expressed in Sartre's trilogy of novels *Roads to Freedom,* and in some of his plays. Later in his life Sartre amended the individualistic existentialism of his early writings and espoused a form of Marxism, which he described as 'the inescapable philosophy of our time,' needing however to be refertilized by existentialism. This change of view is expressed in his *Critique of Dialectical Reason* (Volume 1) of 1960. But I will not attempt to deal with this here; I will consider only the existentialist philosophy of *Being and Nothingness,* making my page references to Hazel Barnes's English translation.

Theory of the Universe

Sartre has some extremely obscure things to say about the nature of 'being' or existence, but for our introductory purposes his most important metaphysical assertion is his denial of the existence of God. He claims that the very idea of God is self-contradictory (p. 615), and does not spend much further argument on the matter. He seems mainly concerned to consider the consequences of God's non-existence for the meaning of our lives. Like Nietzsche, he holds the absence of God to be of the utmost importance for us all; the atheist does not merely differ from the Christian on a point of metaphysics, he must hold a profoundly different view of human existence. If God does not exist, then everything is permitted (as Dostoyevsky once put it). There are no transcendent or objective values set for us, neither laws of God nor Platonic Forms nor anything else. There is no ultimate meaning or purpose inherent in human life; in this sense life is 'absurd.' We are 'forlorn,' 'abandoned' in the world to look after ourselves completely.

Sartre insists that the only foundation for values is human freedom, and that there can be no external or objective justification for the values anyone chooses to adopt (p. 38).

Theory of Man

In one sense, Sartre would deny that there is any such thing as 'human nature' for there to be true or false theories about. This is a typically existentialist rejection of general statements about man. Sartre has expressed it by saying that man's existence precedes his essence (pp. 438–9); we have not been created for any purpose, neither by God nor evolution nor anything else. We simply find ourselves existing, and then have to *decide* what to make of ourselves. Now he can hardly mean to deny that there may be certain properties which are universal among human beings–for instance, the necessity to eat to survive. That there are some such general facts is obvious, although there may be room for dispute about their number. So presumably what he means is that there are no 'true' general statements about what all men *ought* to be, and this is simply the rejection of any notion of objective values, which we have already noted.

Nevertheless Sartre, as an existentialist philosopher, is bound to make *some* general statements about the human condition. His central assertion is of course that of human freedom. We are, in his view, 'condemned to be free'; there is no limit to our freedom except that we are not free to cease being free (p. 439). But we must examine how he reaches this conclusion via an analysis of the notion of consciousness. He starts with a radical distinction between consciousness (*L'être-pour-soi,* being-for-itself) and non-conscious objects (*L'être-en-soi,* being-in-itself; p. xxxix). This basic dualism is shown, he thinks, by the fact that consciousness necessarily has an object; it is always consciousness *of* something which is not itself (p. xxxvii). The next point to appreciate is the connection Sartre sees between consciousness and the mysterious concept of 'nothingness' which appears in the title of his book. We shall be wise to avoid any attempt to trace the roots of this concept in German philosophy, and just pick out some intelligible points from Sartre. We have noted that consciousness is always of something other than itself; Sartre holds that it is always aware of itself as well (pp. xxix, 74–5), so it necessarily distinguishes itself and its object. This is connected with our ability to make judgements about such objects. A judgement can be negative as well as positive; we can recognize and assert what is truly *not* the case, as when I scan the café and say 'Pierre is not here' (pp. 9–10). If we ask a question, we must understand the possibility of the reply being 'No' (p. 5). So conscious beings, by their very nature, can conceive of what is *not* the case.

Sartre makes mystifying verbal play with his concept of nothingness, sometimes in absurdities such as 'the objective existence of a non-being' (p. 5; which, if it means anything, can only mean that there are true negative statements), sometimes in dark sayings like 'Nothingness lies coiled in the heart of being–like a worm' (p. 21). But as far as I can see, the crucial role of nothingness is to make a conceptual connection between consciousness and freedom. For the ability to conceive of what is not the case is the freedom to imagine other possibilities, the freedom to suspend judgement (pp. 24–5). We can never reach a state in which there are no possibilities unfulfilled, for whatever state we are in, we can always conceive of things being otherwise. (Sartre thinks that we are always trying to reach such a state, to become objects rather than conscious beings; hence his description of human life as 'an unhappy consciousness with no possibility of surpassing its unhappy state' (p. 90), 'a useless passion' (p. 615).) The notion of desire involves the recognition of the *lack* of something (p. 87), as does the notion of intentional action (p. 433), for I can only try to achieve a result if I believe that what I intend is not already the case. The power of negation is, then, the same thing as freedom–both freedom of mind (to imagine possibilities) and freedom of action (to try to actualize them). It follows that to be conscious is to be free.

Note how this position of Sartre's directly contradicts two of Freud's. Obviously, it is incompatible with Freud's postulate of complete psychic determinism (p. 459). But it also involves a conflict with the postulate of unconscious mental states, since Sartre holds that consciousness is necessarily transparent to itself. Every aspect of our mental lives is intentional, chosen, and our responsibility. For instance, emotions are often thought to be outside the control of our wills, but Sartre maintains that if I am sad it is only because I choose to make myself sad (p. 61). This view, outlined more fully in his *Sketch for a Theory of the Emotions,* is that emotions are not moods which 'come over us,' but ways in which we apprehend the world. What distinguishes emotion from other ways of being conscious of objects is that it attempts to transform the world by magic—when we cannot reach the bunch of grapes, we dismiss them as 'too green,' attributing this quality to them although we really know quite well that their ripeness does not depend on their reachability. So we are responsible for our emotions, because they are ways in which we choose to react to the world (p. 445). We are equally responsible for longer-lasting features of our character. We cannot just say 'I am shy' as if this were an unchangeable fact about us like 'I am Black,' for our shyness is the way we behave, and we can choose to try to behave differently. Even to say 'I am ugly' or 'I am stupid' is not to assert a fact already in existence, but to anticipate how people will react to my behaviour in future, and this can be discovered only by trying (p. 459).

So even though we are often not aware of it, our freedom and hence our responsibility extend to everything we think and do. There are times, however, when this total freedom is clearly manifested to us. In moments of temptation or indecision—for example, when the man who has resolved not to gamble any more is confronted with the gaming tables once again—we realize that no motive and no past resolution determines what we do *now* (p. 33). Every moment requires a new or renewed choice. Following Kierkegaard, Sartre uses the term 'anguish' to describe this consciousness of one's own freedom (pp. 29, 464). Anguish is not fear of an external object, but the awareness of the ultimate unpredictability of one's own behaviour. The soldier fears injury or death, but feels anguish when he wonders whether he is going to be able to 'hold up' in the coming battle. The person walking on a dangerous cliff-path fears falling, but feels anguish because he knows that there is nothing to stop him throwing himself over (pp. 29–32).

Diagnosis

Anguish, the consciousness of our freedom, is painful, and we typically try to avoid it (p. 40). But such 'escape' is illusory, for it is a necessary truth that we are free. Such is Sartre's diagnosis of man's condition. The crucial concept in his diagnosis is that of self-deception or 'bad faith' (*mauvaise foi*). Bad faith is the attempt to escape anguish by pretending to ourselves that we are not free (p. 44). We try to convince ourselves that our attitudes and actions are determined by our character, our situation, our role in life, or anything other than ourselves. Sartre gives two famous examples of bad faith (pp. 55–60). He pictures a girl sitting with a man who she knows very well would like to seduce her. But when he takes her hand, she tries to avoid the painful necessity of a decision to accept or reject him, by pretending not to notice, leaving her hand in his as if she were not aware of it. She pretends to herself that she is a passive object, a thing, rather than what she really is, a conscious being who is free. The second illustration is of the café waiter who is doing his job just a little too keenly; he is obviously 'acting the part.' If there is bad faith here, it is that he is trying to identify himself completely with the role of waiter, to pretend that this particular role determines his every action and attitude. Whereas the truth is that he has chosen to take on the job, and is free to give it up at any time. He is not *essentially* a waiter, for no man is essentially anything.

Sartre rejects any explanation of bad faith in terms of unconscious mental states. A follower of Freud might try to describe the above cases as examples of repression–in the case of the girl she is repressing the knowledge that her companion has made a sexual advance to her, in the case of the waiter he is repressing the knowledge that he is a free agent who does not have to continue acting as a waiter a moment longer than he wants to. But Sartre points out what seems to be a self-contradiction in the very idea of repression: we must attribute the repressing to some agency within the mind ('the censor') which makes distinctions between what is to be repressed and what is to be allowed into consciousness, so this censor must be aware of the repressed idea in order not to be aware of it. The censor itself is thus in bad faith, and so we have not gained any explanation of how bad faith is possible by merely localizing it in one agency of the mind rather than in the person as a whole (pp. 52–3).

Sartre goes on to suggest that sincerity, the antithesis of bad faith, presents just as much of a conceptual problem. For as soon as we describe ourselves in some way (e.g., 'I am a waiter'), by that very act a distinction is made between the self doing the describing and the self described. The ideal of complete sincerity seems doomed to failure (p. 62), for we can never be just objects to be observed and accurately described. Sartre is expressing here what others have called 'the systematic elusiveness of the self.' But his account makes it even more paradoxical and perplexing than it really is, for he constantly repeats the formula that 'human reality must be what it is not and not be what it is' (e.g., p. 67). This is of course a self-contradiction, so Sartre cannot mean it literally; I think we must take it as shorthand for 'human reality is not *necessarily* what it is, but must be *able* to be what it is not' (my paraphrase of the way he puts it on p. 58). This directs our attention back to Sartre's most basic point, that to be conscious at all is to be free. Consciousness conceals in its being a permanent risk of bad faith, but Sartre maintains that it is *possible* to avoid this and achieve authenticity (p. 70).

Prescription

In view of his rejection of any possibility of objective values, Sartre's prescription has to be a curiously empty one. There is no *particular* course of action or way of life that he can recommend to others. All that he can do is to condemn any bad faith, any attempt to pretend that one is not free. And all that he can recommend is authenticity, that we each make our individual choices with full awareness that nothing determines them for us. We must accept our responsibility for everything about ourselves, not just our actions, but our attitudes, our emotions, and even our characters. The 'spirit of seriousness,' which is the illusion that values are objectively in the world rather than sustained only by human choice, must be repudiated (pp. 580, 626). There is no escape from the anguish of freedom; to flee responsibility is itself a choice (pp. 479, 555–6).

In *Existentialism and Humanism* Sartre illustrates the impossibility of prescription by the case of a young Frenchman at the time of the Nazi occupation, who was faced with the choice of either going to help the free French forces in England or staying at home to be with his mother who lived only for him. One course of action would be directed to what he saw as the national good, but would probably be of insignificant effect in the total war. The other would be of immediate practical effect, but would be directed to the good of only one individual. Sartre holds that no ethical doctrine, Christian or Kantian or any other, can arbitrate between such incommensurate claims. Nor can strength of feeling in the individual faced with the choice settle the matter, for there is no measure of such feeling except in terms of what he actually does, which is of course precisely what is at stake. To choose an adviser is itself a choice. So when Sartre was consulted by this young man, he said merely 'You are free, therefore choose.'

However Sartre does clearly commit himself to the intrinsic value of authentic choice. His descriptions of particular cases of bad faith are not morally neutral, but implicitly condemn any self-deception, any refusal to face reality and admit one's own choices. He thus offers another perspective on the ancient virtue of self-knowledge prescribed by Socrates, Freud, and many others. But Sartre's understanding of the nature and possibility of self-knowledge differs in crucial ways from Freud's. We have seen that psychoanalysis is based on the hypothesis of unconscious mental states which have causal effects on people's mental life. Freud conceived of these causes as acting in a quasi-mechanical way, like flows of energy, and he thought of his job in psychoanalysis as the uncovering of these hidden causes. Sartre emphatically rejects the idea of unconscious causes of mental events; for him everything mental is already out in the open, available to consciousness (p. 571). The job of what he calls 'existential psychoanalysis' is not to look for *causes* of a person's behaviour, but for the *meaning* of it (pp. 568–9). Some contemporary psychiatrists, such as R. D. Laing, follow Sartre on this point. (In the last chapter I mentioned the suggestion that Freudian psychoanalysis itself is really an interpretation of motives, purposes, and intentions rather than a discovery of causes.) So to understand a person Sartre looks for *choices* (p. 573), and he holds that since a person is essentially a unity, not just a bundle of unrelated desires or habits, there must be for each a person a fundamental choice (the 'original project') which gives the meaning of every particular aspect of his behaviour (pp. 561–5). The biographies Sartre has written of Baudelaire, Genet, and Flaubert are particular exercises in interpreting the fundamental meaning of a person's way of life. Existential psychoanalysis is, then, the means by which Sartre hopes that we can achieve genuine self-knowledge. He ends *Being and Nothingness* with a promise to write another work, on the ethical plane, to show how we can live as free beings aware of our freedom.

Critical Discussion

My first complaint against Sartre is one about style rather than content. *Being and Nothingness,* it is only fair to warn the reader, is easily the most unreadable of the texts I refer to in this book. This is a matter not just of length and repetitiousness, but of a word-spinning delight in the abstract noun, the elusive metaphor, and the unresolved paradox. To trace this to the influence of Hegel, Husserl, and Heidegger may explain but can hardly excuse it. One may be thankful that Sartre is not as obscure as they are, but surely he could have said what he had to say more clearly and a lot more briefly. It is all the more tantalizing when one finds passages of relative clarity and great insight buried inside the conglomeration of verbiage.

However the effort to understand him does begin to reveal a view of human nature which has a certain compelling fascination.

To turn to matters of content, let us first consider the problem of how bad faith is possible. We have noted that Sartre rejects, for conceptual reasons, any Freudian explanation of this. But it is not clear whether he offers any adequate solution of his own to the conceptual problem of how consciousness can 'be what it is not and not be what it is,' despite the extended discussion of being-for-itself in Part Two of *Being and Nothingness*. He appears to rest too easily in such paradoxical statements and to shirk the difficult philosophical task of explaining in clear unparadoxical terms what it is about consciousness that generates the problem.

There is an apparent contradiction between Sartre's constant insistence on our freedom, and his analysis of the human condition as necessarily determined in certain respects. For he holds that as conscious beings we are always wanting to fill the 'nothingness' which is the essence of our being conscious; we want to become things rather than remain perpetually in the state of having possibilities unfulfilled (p. 90). He also holds that the relationship between two consciousnesses is necessarily one of conflict, for each wants to achieve the impossible ideal of making the other into a mere object (pp. 394, 429). In these two respects, he analyses human life as a perpetual attempt to achieve the logically impossible. But why *must* it be so? Surely there is a direct contradiction between these 'musts' and our supposed freedom? Cannot someone *choose* not to want to become an object, or to make other people into objects? It is hard to see whether Sartre even tries to resolve these tensions at the heart of his theory.

We have noted that the only positive recommendation that Sartre can make is that one should avoid bad faith and choose authentically. But can self-knowledge or authenticity be the only basis for how to live? If no reasons whatsoever can be given for choosing one way of life rather than another, the choice is arbitrary. It looks as if on his own premisses Sartre would have to commend the man who chooses to devote his life to exterminating Jews, provided that he chooses this with full awareness of what he is doing. Conversely, the man who devotes himself to helping 'down and outs' but will not avow his own real motive for doing so (perhaps a reaction against his parental background) would apparently have to be condemned as inauthentic. Or can it be argued that authenticity must involve respecting the freedom of other people? Sartre never wrote his promised book on ethics, and perhaps the reason is that he realized that no social ethic could be developed from the individualist premisses of *Being and Nothingness*. No doubt this is also why he came to adopt a Marxist standpoint, to seek the social conditions which would make it possible for *all* men to exercise their freedom.

Yet there is something important to learn from Sartre's deep analysis of how the very notion of consciousness involves that of freedom. We have seen how he wants to extend the concept of choice far beyond its normal use, to hold us responsible not just for our actions, but for our emotions and even for our characters. If I am angry, it is because I have chosen to be angry; and if I am the kind of person who is usually passively resigned to his condition, that too is a disposition that I choose to adopt. This view does seem to contradict our normal concepts of emotion and character, for emotions are assumed to 'come over one' whether one wills it or not. And our character is supposed to be a *fact* about us like our weight–something we can try to change gradually by taking certain steps, admittedly, but not something which we can change at a stroke, like standing up or sitting down. And yet Sartre's view here is not just an arbitrary misuse of language. For we do commonly reproach people for their emotions and characters–'How *could* you feel like that?' '*Must* you be so . . . ?' And such reproach is not always useless. For to make someone *aware* that he is feeling or behaving in a certain way does make a difference to him. The more he is aware of his anger or pride, the more he is not *just* angry or proud, and the more capable he is of becoming something else. Perhaps this is the essence of Sartre's point.

The vast verbiage of his philosophy issues ultimately in a directly practical and intimate challenge to us all, to become more truly self-aware and to exercise our power of changing ourselves.

For Further Reading

Basic text: *Being and Nothingness,* translated by Hazel Barnes (Methuen, London, 1957; Citadel Press paperback, Secaucus, N.J.). This lengthy and difficult book is best read with the help of an introduction which directs one to the more important and relatively clear passages—I have excerpted some of these in *The Study of Human Nature* (Oxford University Press, New York, 1981).

Sartre, by A. C. Danto (London, Fontana, 1975) is a brief and stimulating attempt to interpret *Being and Nothingness* in the light of analytical philosophy. *Sartre's Concept of a Person: An Analytic Approach,* by Phyllis Sutton Morris (University of Massachusetts Press, Amherst, 1976) is another useful work in the same vein.

Sartre: A Philosophic Study, by Anthony Manser (Athlone Press, London, 1966; Greenwood, Westport, Conn., 1981) is a survey of the whole of Sartre's thought, including his literature and politics but giving most attention to his philosophy.

For an introduction to other existentialists, see *Existentialism,* by Mary Warnock (Oxford University Press OPUS paperback, Oxford, 1970).

8

Skinner: The Conditioning of Behaviour

Perhaps the reader is wondering by now whether it is worth giving so much attention to the philosophers and speculative thinkers of the past. In a scientific age, should we not look to psychology for the truth about human nature? In the last hundred years psychology has established itself as an independent branch of empirical science clearly separated from its early philosophical sources, so surely we can now expect some properly scientific answers to our questions about human nature? The fact is, however, that psychology is an extraordinarily difficult and complex discipline, which yields clear answers only to very carefully and precisely defined questions on specific topics. So if an experimental psychologist begins to *generalize* about human nature, his statements are likely to be as speculative as those of the thinkers we have considered, at least in the present state of his subject. It is also true that there are still various schools of thought and methodologies within psychology, so that it is not as free from 'philosophical' questions as we might like to think. As a sample of what one kind of psychology has to offer, let us consider the work of B. F. Skinner, Professor of Psychology at Harvard University 1948–1974, who has been one of the most influential experimental psychologists in the behaviourist tradition. Since he is also one of those who is prepared to generalize about human nature and to offer diagnoses of, and prescriptions for, our problems, we will find plenty to discuss without necessarily being drawn into the details of his experimental work, upon which non-psychologists can hardly be qualified to comment.

As background to Skinner's work it will be useful to consider that of his earlier compatriot J. B. Watson, who is generally recognized as the founder of psychological behaviourism. In the last quarter of the nineteenth century psychology began to be an empirical science, and the first psychological laboratories had been set up under the leadership of men like Wundt in Germany and William James in America. They defined it not as the study of soul or mind (which would imply some sort of metaphysical dualism) but as the study of consciousness. They thought that since each of us is aware of the contents of his own consciousness, we can simply report them by introspection and thus give the empirical data for psychology. But it was soon found that such reports could not agree on the description and classification of sensations, images, and feelings. So the introspective method ran into an impasse. At the same time Freud's work was suggesting that important aspects of the mind were not accessible to consciousness. In the study of animals introspection is obviously not available, and yet (since Darwin) one would expect the study of animals to be closely related to that of men. And in any case the notion of consciousness poses almost as many philosophical problems as that of soul or mind.

So when Watson proclaimed, in a paper of 1913, that the subject matter of psychology should be *behaviour,* not consciousness, his views found a ready acceptance which reorientated psychology completely. For the behaviour of animals and men is publicly observable, so reports and descriptions of it can form the objective data of psychology; and the concept of behaviour apparently involved no questionable philosophical assumptions. This rejection of the introspective method was the most fundamental point of Watson's new programme. It is of course a purely *methodological* dictum about what psychology ought to study, and is quite independent of any metaphysical statement that consciousness does not exist, or that it is nothing but the material processes inside a person's skull. It is also independent of the philosophical thesis (called logical or analytical behaviourism) that our words for mental phenomena really refer only to behaviour and dispositions to behaviour. But Watson and his followers were wont to go beyond their merely methodological point and allege that belief in consciousness is a hangover from our superstitious pre-scientific past, akin to belief in witchcraft.

There were two other main points in Watson's creed, however, which are really empirical theories within psychology rather than methodological points. The first was his belief that environment is much more important than heredity in the determination of behaviour. This is a natural concomitant of his methodology, for the external influences on an organism's behaviour are easily observable, and manipulable by experiment, whereas the internal influences (and in particular the genes) are much more difficult to observe and manipulate. But of course this fact alone does not tell us anything about the relative influence of environment and heredity on behaviour. However Watson assumed that the only inherited features of behaviour were simple physiological reflexes; he attributed everything else to learning. Hence his claim (which he admitted went beyond the known facts): 'Give me a dozen healthy infants, well-formed, and my own specified world to bring them up in and I'll guarantee to take any one at random and train him to become any kind of specialist I might select–doctor, lawyer, artist, merchant-chief, and yes even beggar-man and thief, regardless of his talents, penchants, abilities, vocations, and race of his ancestors' (*Behaviourism,* 1924, revd. edn. 1930, p. 104). Watson's other empirical guess was a particular theory of how learning takes place, namely by the conditioning of reflexes. This was suggested by Pavlov's famous experiments in which he trained dogs to salivate at the sound of a bell, by regularly ringing the bell just before feeding them. Watson's programme was to explain all the complex behaviour of animals and men as the result of such conditioning by their environment.

In the work of experimental psychology since Watson's time, doubt has been cast both on his extreme emphasis on environment, and on his particular theory of learning. However, Skinner has carried on with Watson's programme. He sticks to the behaviourist methodology even more rigorously, and eschews all reference to unobservable entities. He shows a similar faith in the programme of explaining all behaviour of animals and men as the effect of the environment upon them, an effect mediated by a few basic conditioning processes. *The Behaviour of Organisms: An Experimental Analysis* (1938) is his fundamental technical work on conditioning. In *Science and Human Behaviour* (1953) he applied his theories to human life and society in general, and in *Verbal Behaviour* (1957) to human language in particular. He has also published a novel, *Walden Two* (1948), in which he describes a Utopian community organized on his principles of behavioural conditioning. And recently he has produced *Beyond Freedom and Dignity* (1971) in which he claims again that a technology of behaviour can solve the problems of human life and society, if only we will give up our illusions about individual freedom, responsibility, and dignity. In what follows I shall make my page references to *Science and Human Behaviour,* which is the most wide-ranging and readable of these works. And I shall incorporate criticism with exposition, because the analysis of what Skinner means leads directly to criticism of it.

Theory of the Universe

Skinner is the most rigorously 'scientific' of the thinkers I consider in this book. He believes that only science can tell us the truth about nature, including human nature, for science is unique in human activity in showing a cumulative progress (p. 11). What is fundamental to science is neither instruments nor measurement but scientific *method*–the disposition to get at the facts, whether expected or surprising, pleasant or repugnant. All statements must be submitted to the test of observation or experiment, and where there is insufficient evidence we must admit our ignorance. The scientist tries to find uniformities or lawful relations between phenomena, and to construct general theories which will successfully explain all particular cases (pp. 13–14). Furthermore Skinner sees no clear distinction between science and technology; he says that the job of science is not just to predict, but to *control* the world (p. 14).

Most scientists would agree with Skinner in his description of scientific method, except that they might make a clearer distinction between science and technology, between prediction and control. But some scientists are Christians while others are humanists, some are left-wing and others right-wing. Skinner seems to think that there is no basis except in science for answering *any* sort of question. Certainly, he finds no scientific basis for belief in God, and treats religion as merely one of the social institutions for manipulating human behaviour (pp. 350–8). Value judgements are, he thinks, typically the expression of the pressure to conform which is exerted by any social group (pp. 415–18), a kind of concealed command (p. 429). They can be given an objective scientific basis only if they concern means to ends. 'You ought to take an umbrella' might be roughly translated as 'You want to keep dry, umbrellas keep you dry in the rain, and it's going to rain' (though Skinner replaces the ordinary notion of wanting by the supposedly more scientific notion of 'reinforcement'; p. 429). The only objective basis he can see for evaluating cultural practices as a whole is their survival value for the culture (pp. 430–6). But even here he says that we do not really *choose* survival as a basic value, it is just that our past has so conditioned us that we do tend to seek the survival of our culture (p. 433). If we want a label for this attempt to answer *all* questions purely scientifically, perhaps 'scientism' will do.

Theory of Man

Skinner proposes that the empirical study of human *behaviour* is the only way to arrive at a true theory of human nature. So naturally he will reject any kind of metaphysical dualism. But with it he rejects any attempt to explain human behaviour in terms of mental entities, whether they are everyday concepts of desires, intentions, and decisions, or Freudian postulations such as id, ego, and superego (pp. 29–30). He rejects such entities not only because they are unobservable, but because he thinks they are of no explanatory value anyway. For instance, to say that a man eats because he is hungry is not to assign a cause to his behaviour but simply to redescribe it (p. 31). It is no more explanatory than saying that opium puts you to sleep *because* it has a 'dormitive power.' Of course, Skinner must admit the possibility of discovering physiological preconditions of behaviour (the literally inner states). But he holds that even when the progress of physiology tells us about these, we shall still have to trace *their* causation back to the environment, so we may as well bypass the physiology and look directly for the environmental causes of behaviour (p. 35). He has to admit that genetic factors are relevant, for it is obvious that different species behave in very different ways. But he dismisses the layman's use of 'heredity' as a purely fictional explanation of behaviour, and holds genetic factors to be of little value in 'experimental analysis' because they cannot be manipulated by the experimenter (p. 26).

This position is a rather confusing mixture of methodological precept and empirical theory, both derived from Watson's behaviourism. We must try to sort out the different components in the mixture. Obviously Skinner defines psychology as the study of behaviour. But this does not settle whether psychology is to be permitted to postulate unobservable entities to explain behaviour. Many psychologists are quite happy to talk in terms of drives, memory, emotions, and other 'mental' entities, provided of course that what is said about them is testable by the observation of behaviour. But it appears that Skinner is adopting a much more austere methodology, and is rejecting all mention of unobservables. In this he is trying to be more 'scientific' than most scientists, for the physical sciences very often postulate unobservable theoretical entities such as magnetic fields, mechanical forces, and subatomic particles. In the heyday of the philosophy of logical positivism it was doubted whether this was really proper procedure, but it is now generally acknowledged that to disallow it would be an impossible restriction on scientific method. Provided that what is said about unobservable theoretical entities is falsifiable by observation, there is no valid objection to them. So if Skinner is rejecting inner mental causes of behaviour *just* because they are unobservable, then I think we must judge this to be a quite unnecessarily restrictive methodology.

But he does offer another reason for rejecting what he calls 'conceptual' inner causes (p. 31), namely that they are of no explanatory value (as noted above). However he has not shown that such conceptual inner causes must be merely redescriptive of what they are supposed to explain; he has only given a few examples in which he thinks this to be true. Certainly, an inner state S can only be a genuine explanation of behaviour B if we can have some evidence for the existence of S other than the occurrence of B, but surely this condition is sometimes satisfied. For instance (in Skinner's own example) we can have good evidence for saying that someone is hungry even though he is not *actually* eating, if we know that he has not eaten for 24 hours (and perhaps he *says* he is hungry!). It is just not true that a single set of facts is described by the two statements: 'He eats' and 'He is hungry.' Obviously, one can be hungry when one is not eating, and (less often) one can eat when one is not hungry. Skinner has not given any adequate reason for the rejection of all conceptual causes of behaviour.

What of his rejection of physiological states as causes? The fact that these are not easily observable or manipulable does nothing to show that they do not play a crucial role in the causation of behaviour (as we noted when discussing Watson's environmentalism). Skinner's assumption is that physiological states inside an organism merely mediate the effect of its environment (past and present) on its behaviour. So he thinks that psychology can confine its attention to the laws connecting environmental influences directly with behaviour. There are two separable assumptions here. Firstly, that human behaviour is governed by scientific laws of *some* kind: 'If we are to use the methods of science in the field of human affairs, we must assume that behaviour is lawful and determined' (pp. 6 and 447). Secondly, that these laws state causal connections between environmental factors and human behaviour: 'Our "independent variables"—the causes of behaviour—are the external conditions of which behaviour is a function' (p. 35).

These two assumptions could be taken in a purely methodological interpretation, as expressing a programme of *looking* for laws governing human behaviour, and specifically for laws connecting environment with behaviour. As such, there can be no objection to them. But it is fairly clear that Skinner also takes them as general assertions of what is the case about human behaviour. As such, we must ask whether there is any good reason to think that they are true, for these are the crucial assumptions on which Skinner's theory of human nature is based. (They are recognizably descended from Watson's environmentalism.) Firstly, do we have to assume that *all* human behaviour is governed by causal laws, if we are to study that behaviour scientifically? There is no more reason to assume this than there is for Marx to maintain that if we are to study history scientifically there must be laws which determine everything which happens. Universal determinism is not a necessary presupposition of science, although the *search* for causal laws is central to science. Admittedly, it would be rather disappointing if psychology could not proceed beyond the mere reporting of particular events and statistical regularities. But whether there are causal laws governing behaviour is something which we must leave psychology to discover. That *all* behaviour is governed by such laws is a 'metaphysical' assumption which ill befits a supposedly strict empiricist such as Skinner.

The more specific assumption that all behaviour is a function of *environmental* variables is even more dubious. What it means, in detail, is that for any piece of behaviour, there is a finite set of environmental conditions (past or present) such that it is a causal law that anyone to whom all those conditions apply will perform that behaviour. This is reminiscent of Watson's claim that he could take any infant at random and make of it anything he liked, given only the appropriate environment. It entails a denial that inherited factors make any difference to the behaviour of human beings. So that, for instance, any healthy infant could be trained to become a four-minute miler, a nuclear physicist, or anything else. In its full generality, this claim seems fairly obviously false. The fact that the differences in ability between identical twins brought up apart are much less than the average range of ability in the whole population is evidence against it. Heredity does play *some* part, though this is not to deny the huge importance of environment. To attribute all to environment is another assumption which Skinner does not submit to empirical test.

We should now pay some brief attention to the specific mechanisms of conditioning by which Skinner thinks the environment controls behaviour. His theory is descended from the ideas of Pavlov and Watson, but this is the area in which Skinner has made his own major contributions to the advancement of psychological knowledge. In the 'classical' conditioning of Pavlov's experiments, the 'reinforcer' (food) was repeatedly presented together with a 'stimulus' (the ringing of a bell), and the 'response' (salivation) would then appear for the bell without the food. The main difference in Skinner's 'operant' conditioning is that what is conditioned is not a reflex response like salivation, but any kind of behaviour which the animal may perform quite spontaneously without any particular stimulus. For instance, rats can be trained to press levers, and pigeons to hold their heads abnormally high, in each case simply by feeding the animal whenever it presses the lever, or raises its head above a certain level. So when the environment is arranged such that the reinforcer follows upon a certain kind of behaviour (called the 'operant,' since the animal thus operates on its environment), then that behaviour is performed more frequently (pp. 62–6). (This is of course the general principle on which all animal training works.) In a vast amount of careful experimental work, Skinner and his followers have discovered many new facts about the processes of conditioning, for instance, that intermittent reinforcement tends to produce a greater rate of response–so if we want a rat to do as much lever-pressing as possible, we should feed it irregularly, not after every press.

Skinner's experimental work is impressive and unimpugnable, but what we can and must criticize here is his extrapolation from it to human behaviour in general. In *Science and Human Behaviour* he outlines the understanding of behaviour that he has gained from his animal experiments (mainly with rats and pigeons) and then goes on to apply these conceptions to human individuals and institutions–government, religion, psychotherapy, economics, and education. But it is quite possible that Skinner's discoveries about rats and pigeons apply only to those species (and perhaps related ones), but not to more complex animals and especially not to men. Although he rightly points out that we cannot assume that human behaviour is different in kind from animal behaviour (pp. 38–9), his whole approach seems to make the equally unjustified assumption that what applies to laboratory animals will apply (with only a difference of complexity) to men (pp. 205 ff.).

One very important area in which Skinner has applied his theories to human behaviour is that of language. In *Verbal Behaviour* he attempts to show that all human speech can be attributed to the conditioning of speakers by their environment (which includes of course their social environment, the noises made by surrounding humans). Thus, a baby born in England is subjected to many samples of English conversation, and when its responses are reasonably accurate reproductions of what it has heard, they are 'reinforced,' and thus the child comes to learn to speak English. Adult speech, too, is analysed by Skinner as a series of responses to stimuli from the environment, including verbal stimuli from other people.

The crucial defects in Skinner's account of language have been pointed out by Chomsky, whose work has given new direction to research in linguistics and psychology in the last decade. Chomsky argues that, although Skinner has tried to describe *how* language is learned, his account is of little value because he pays no attention to the question of *what* it is that we learn when we acquire the ability to speak a language as our native tongue. Clearly, we can hardly ask how we learn X unless we first know what X is; we must have a criterion for someone having *succeeded* in learning X. Now human language is a very different sort of phenomenon from rats' pressing of levers. Skinner could hardly deny this, but would suggest that the differences are only a matter of degree of complexity. But Chomsky suggests that the *creative* and *structural* features of human language–the way in which we can all speak and understand sentences we have never heard before, just by our knowledge of the vocabulary and grammar of our language—make it quite different in kind from any known kind of animal behaviour. If so, the attempt to analyse human speech in terms derived from the behaviour of lower animals would seem to be doomed from the start. And the same would apply to other distinctively human forms of behaviour.

Even the suggestions Skinner does make for how linguistic behaviour is learned can be seen to be based on very shaky analogies. For instance, the reinforcement which may encourage correct speech by an infant is very rarely feeding, but more likely some sort of social approval. He suggests that we can be reinforced by being paid attention, or even by merely saying something which is satisfying to ourselves, perhaps just because it is an accurate reproduction of what we have heard. The trouble here is that this is merely speculation. The use of a term like 'reinforcement' which has a strictly defined meaning for certain experiments with animals in no way guarantees scientific objectivity for its use in human situations which are allegedly analogous. So once again, Skinner's supposedly strictly empirical approach turns out to conceal a large element of unempirical speculation.

There is another important respect in which Chomsky argues that Skinner's theories fall down when applied to human language. This is the matter of inherited factors, of the contribution made by the speaker rather than his environment to his learning of language. Obviously, English children learn English, and Chinese children learn Chinese, so the environment does have major effects. But again, all normal human children learn one of the human languages, but no other animal learns anything which resembles human languages in the crucial respect of the formation of indefinitely many complex sentences according to rules of grammar. So it seems that the capacity to learn such a language is peculiar to the human species. Skinner holds that our language-learning must be due just to a complex set of reinforcements from our human environment. Chomsky suggests that the amazing speed with which children learn the grammatical rules of the language they hear from a very limited and imperfect sample of that language can be explained only by the assumption that there is in the human species an *innate* capacity to process language according to such rules. So behind all the apparent variety of human languages there must be a certain basic systematic structure common to all, and we must suppose that we do not *learn* this structure from our environment, but process whatever linguistic stimulation we receive in terms of this structure. This fascinating hypothesis has by no means been proved, but the available evidence does tend to favour it rather than Skinner's extreme environmentalism.

Speech is of course not the only human activity. But it is especially important as a representative of the 'higher' human mental abilities. So if Skinner's theories fail to explain it adequately we must conclude that even if they explain some human behaviour they cannot give a true account of human nature in general. There remains the possibility that other important aspects of human behaviour are not learned from the environment but are genuinely innate.

Diagnosis

Skinner's diagnosis can be seen as the exact opposite of Sartre's. Sartre maintains that we are free, but keep pretending that we are not. Skinner says we are determined, but still like to think that we are free. He analyses our current social practices as based on theoretical confusion. We are increasingly realizing how environment determines behaviour, and hence we exonerate people from blame by pointing to the circumstances of their upbringing. Yet we also maintain that people are often genuinely responsible for their actions (p. 8). We are thus in an unstable transitional stage, and 'the present unhappy condition of the world may in large measure be traced to our vacillation'; 'we shall almost certainly remain ineffective in solving these problems until we adopt a consistent point of view' (p. 9). 'A sweeping revision of the concept of responsibility is needed' (p. 241), for our present practice of punishment is remarkably inefficient in controlling behaviour (p. 342). We will have to abandon the illusion that men are free agents, in control of their own behaviour, for whether we like it or not we are all 'controlled' (p. 438).

This diagnosis of 'the unhappy condition of the world' seems very dubious. Admittedly there are important practical problems about deciding the extent of responsibility, and these are closely connected with deep theoretical and philosophical questions about the concept of freedom. But Skinner's dismissal of the concept is an inadequate and unargued response to these problems. In his book, *Beyond Freedom and Dignity,* he seems to be saying that just as it was the mistake of animism to treat inanimate *things* as if they were people and attribute thoughts and intentions to them, so it is a mistake to treat *people* as people and attribute desires and decisions to them! Of course this is absurd. One preliminary point towards getting us out of this confusion is as follows. The thesis of universal determinism is that every event (including all human choices) has a set of sufficient preceding causes. Now even if this thesis is true (and remember that Skinner has given us no reason to believe it), we are not precluded from picking out as 'free' those human actions which include among their causes the *choice* of the person. The concept of a free action surely does not imply that it has no causes at all (that would make it random), but that it is a result of the agent's choice. We might still hold people responsible for the actions they choose, even if those choices themselves have causes. Skinner himself seems to believe it important to use methods of social control which depend on individual choice rather than on forms of conditioning of which people are not aware.

Prescription

Like Marx, Skinner holds that human circumstances can and should be humanly formed. If the environment makes us what we are, then we should change the social environment deliberately so that the human product will meet more acceptable specifications' (p. 427). He thinks that psychology has reached the point where it can offer techniques for the manipulation and control of human behaviour, hence for changing human society for better or for worse (p. 437). If we will only give up the illusions of individual freedom and dignity, we can create a happier life by conditioning everyone's behaviour in appropriate ways. For instance, we can give up the inefficient practice of punishment, and instead induce people to act legally by making them *want* to conform to the standards of society (p. 345). This can be done by a combination of education and positive inducements, not necessarily by propaganda or any concealed manipulation. Thus science could lead to the design of a government which will really promote the well-being of the governed (p. 443), and perhaps even to a set of 'moral values' (Skinner's quotes!) which may be generally accepted (p. 445). Provided that control is diversified between different individuals and institutions there need be no danger of despotism (pp. 440–6).

This vague programme sounds naively optimistic and yet sinister in its jaunty dismissal of individual freedom. What Skinner has in mind comes out a little more clearly in his novel *Walden Two,* in which his ideal community combines the culture-vulture atmosphere of an adult-education summer school with the political system of Plato's *Republic* (for there is a wise designer of the community who has arranged everything on 'correct' behaviourist principles from the start!). But Skinner's utopia is open to much the same objections as Plato's. On what basis are the designers of a culture to decide what is best for everyone? How can misuse of their power be prevented? Despite his mention of safeguards against despotism, Skinner seems politically very naive. His very terminology of 'designing a culture' and 'the human product' suggests that he makes the highly questionable assumption that it should be the aim of social reform to produce a certain ideal kind of society and individual. An important alternative view is that the aim should be purely negative–to eliminate specific causes of human unhappiness, such as poverty, disease, and injustice–and that to try to condition people according to some blueprint is to trespass upon what should be the area of individual choice. (This is the distinction which Popper makes, in his criticism of Plato noted above in Chapter 3, between 'utopian' and 'piecemeal' social engineering.)

So we do not have to accept Skinner's judgement that individual freedom is a myth and therefore not important. There are immediate practical issues involved here, for behaviour therapy based on Skinnerian principles of conditioning has already been applied to neurotics and criminals in some places. But in cases not of physical illness but of behaviour which is 'abnormal' or 'deviant' by some criterion, when (if ever) does anyone have the right to try to condition someone else's behaviour? There are deep problems–factual, conceptual, and ethical–about how the purely scientific approach to a person, as an organism whose behaviour has identifiable and manipulable causes, can be combined with the ordinary assumption by which we treat our fellows as rational beings who are responsible for their intentional actions. Skinner assumes that the two are simply incompatible, and that the latter must give way to the former (p. 449). But this is just the dogmatic and uncritical position taken by one particular psychologist. It would be a great pity if this discouraged us from seeking better understanding of human nature from experimental psychology.

For Further Reading

Basic text: B. F. Skinner, *Science and Human Behaviour* (Macmillan, New York, 1953; Free Press paperback, New York, 1965).

Skinner's Utopian novel *Walden Two* (Macmillan, New York, 1948) and his *Beyond Freedom and Dignity* (Penguin, London, 1973; Bantam Books paperback, New York, 1972) outline his ideal society and the means by which he thinks we can achieve it.

Beyond the Punitive Society: Operant Conditioning and Political Aspects, edited by Harvey Wheeler (Wildwood House, London, 1973) is a collection of critical evaluations of Skinner's ideas.

For a review of the progress of experimental psychology since Watson, see *Behaviour* by D. E. Broadbent (Methuen, London, University Paperbacks 1961). This is written for the layman, and has some critical discussion of Skinner's work in Chapter 5. For a wider-ranging, historically minded introduction to psychology, see G. A. Miller and R. Buckout, *Psychology: The Science of Mental Life* (Harper & Row, New York, 2nd edn. 1973).

For an introduction to Chomsky's views, see his *Language and Mind* (Harcourt Brace Jovanovich, New York, enlarged edition 1972), or the book on Chomsky by J. Lyons in the Modern Masters Series (Fontana, London, 1970; Viking, New York, 1970).

In *The Study of Human Nature* (Oxford University Press, New York, 1981) I have included many readings which are very relevant to the themes of this chapter especially those from Hume, Mill, Watson, and Chomsky in Part III ('The Search for a Scientific Theory of Human Nature') and those in Part IV ('How Far Is a Scientific Theory of Human Nature Possible?').

Psychology has now come a long way from Skinnerian behaviourism; in what is presently called 'cognitive science,' mental faculties and processes are freely postulated, in the hope that such theorizing can be empirically tested by indirect methods. For example, *The Modularity of Mind,* by Jerry A. Fodor (MIT Press, Cambridge, Mass., 1983) shows just how much the subject has changed.

9

Lorenz: Innate Aggression

We have criticized Skinner for neglecting the possibility that certain important features of human behaviour are innate rather than learned from experience. So let us now turn to Lorenz, who bases his diagnosis of human ills on precisely this possibility. Lorenz is one of the founding fathers of the branch of the life sciences called ethology. Etymologically, the term 'ethology' means the study of character, but it is now used to mean the scientific study of animal behaviour. However this does not make clear how an ethologist differs from a psychologist, who would also claim to study animal behaviour scientifically. Perhaps ultimately there is no difference, but there certainly have been two different approaches which are only now achieving some *rapprochement*. We have seen that behaviourists such as Watson and Skinner have been committed not only to the methodology of studying behaviour rather than consciousness, but also to certain wide-ranging empirical theories—that behaviour is caused almost entirely by environmental influences mediated through conditioning mechanisms. Consequently, their experiments have studied the details of how environment can *change* behaviour. In the twenties and thirties of this century, the early ethologists realized that very many of the behaviour patterns of animals (those which have traditionally been called 'instinctual') could not be explained in the behaviourists' way. What was distinctive of such behaviour was that it was *fixed,* it could not be eliminated or altered by the environment, however much that environment was experimentally manipulated. So the ethologists concentrated on these fixed 'instinctual' behaviour patterns, and observed the 'natural' behaviour of the animal in the wild before intervening to perform experiments.

So the distinctive emphasis of ethology arises out of the anti-behaviourist assumption that some of the most important aspects of animal behaviour are innate. To explain such behaviour ethologists appeal not to the past experience of the *individual* animal, but to the process of evolution which has given rise to that *species* of animal. To account for an instinctive behaviour pattern in a species we must say what survival value it has for the genes for that behaviour. So ethology is based, more directly than psychology in general, on the theory of evolution. And contemporary ethological theories of human nature appeal to the evolutionary past of man in order to explain his present condition. This therefore seems the appropriate place to sketch briefly the essentials of the theory of evolution, which is in any case something that no adequate theory of human nature can afford to neglect.

Darwin was not the only person to arrive at a theory of the evolution of species by gradual divergence from common ancestors, but his *Origin of Species* (1859) is the classic work which convinced both scientific and popular opinion of the truth of the theory. Its full title is 'The origin of species by means of natural selection: or the preservation of favoured races in the struggle of life,' which effectively summarizes its key idea. It was written for the general educated public, and documents the argument with an immense wealth of evidence which Darwin had accumulated from his research during the previous twenty years. In this book he does not explicitly state what it clearly implied–that man too was descended from animal ancestors–but this obvious implication caused a famous controversy with some of the theologians of the time. In later books Darwin did explicitly apply his theories to man, in *The Descent of Man* (1871) and *The Expression of the Emotions in Man and Animals* (1872) (note how the title of the latter suggests an ethological theme).

Darwin's theory is a logical deduction from four true empirical propositions. The first two concern matters of genetics—that traits of parents tend in general to be passed on to their offspring, but there is nevertheless considerable variation between individuals of a given species. These two truths emerge clearly from a wide variety of observations, and they are utilized in the deliberate breeding of different varieties of domestic animals. But their theoretical explanation was not discovered until after Darwin's time, in Mendel's theory of genes. (The facts of 'mutation'-spontaneous changes in genes, the number of genes available, and their patterns of interaction, explain the variation stated in the second proposition.) The third and fourth premisses of Darwin's argument are the facts that species are capable of a geometric rate of increase, whereas the resources of the environment cannot support such a rate. It follows from these last two facts that a very small proportion of seeds, eggs, and young reach maturity; in short, there is a struggle for existence, primarily between members of the same species. Now from this struggle and from the fact of variation within a species, we can deduce that there will be certain individuals (those whose characteristics are most 'advantageous' in the given environment) which will live longest. They will have the best chance of leaving offspring; therefore, given the first fact of inheritance, their traits will tend to be passed on, whereas disadvantageous traits will tend to die out. Thus over a period of time the typical characteristics of a population of animals can change. And given the immense periods of geological time, and the wide variety of environments, different species can slowly evolve from common ancestors. All that is needed to produce such evolution is the constant pressure of natural selection acting on the variations caused by random mutations. There is no need to postulate the biologically implausible inheritance of 'acquired' (individually learned) characteristics, as Lamarck did, and as Darwin himself did at some stages of his work.

Apart from this very general argument for the mechanism of evolution, there is much direct empirical evidence for man's common ancestry with other animals. Comparative anatomy shows the human body to have the same general plan as other vertebrates–four limbs with five digits on each, etc. The human embryo goes through stages of development in which it resembles those of the various lower forms of life. In the adult human body there are 'remnants' of such lower forms–e.g., a vestigial tail. The basic chemistry of our bodies–e.g., digestion, blood, genes–is similar to those of other mammals. Finally there are the fossil remains of creatures which were ape-like but resembled humans more than any existing apes. So our animal ancestry is overwhelmingly confirmed by the evidence. Some questions may remain about the detailed mechanism of evolution, but that we have evolved is now an established fact, which no true theory of human nature can contradict.

But exactly what implications the fact of evolution has is a matter of dispute which cannot be settled by the theory itself. Some nineteenth-century churchmen thought it contradicted the Christian doctrine of creation, but most present-day theologians find no real conflict (Teilhard de Chardin has even constructed a peculiarly evolutionary theology). Marx welcomed the theory as a confirmation of his view of the progressive development of human history (he even wanted to dedicate the English edition of *Das Kapital* to Darwin, but the latter politely declined the honour!). Yet right-wing politicians claimed that evolution showed unrestrained economic competition to be 'natural,' like the survival of the fittest, and therefore right (such doctrine was called 'social Darwinism'). In our own day several popular books have used the idea that our evolution from ape-like ancestors is the key to our true nature: Robert Ardrey in *The Territorial Imperative* and other books, Desmond Morris in *The Naked Ape,* and Arthur Koestler in *The Ghost in the Machine.* But one of the main sources of their ideas is Lorenz's work. So a critical examination of his ideas should equip the reader to look at these recent popular works with a sympathetic but sceptical eye.

Like Freud, Konrad Lorenz is a product of the scientific and cultural traditions of Vienna who has pioneered a new area of scientific study with deep implications for mankind. In his technical papers on animal behaviour he has interpreted his very extensive and careful observations of many species, and some of the concepts he introduced have passed into the common currency of modern biological science. But he has also written for the general reader, and in *King Solomon's Ring* (1950), *Man Meets Dog* (1954), and *On Aggression* (1963), he displays style, humour, an engaging personality, and an awareness of deep issues of epistemology and society. The first two books introduce ethological themes by a variety of anecdotal descriptions, mostly of the pets Lorenz himself has kept. The latter concentrates on aggressive behaviour and attempts a diagnosis of man's condition, so my page references will be to it (for the American edition see the end of this chapter).

Theory of the Universe

Lorenz is a biological scientist, so the most important of his background assumptions is the theory of evolution which I have summarized above. To explain the existence of any particular organ or behaviour pattern he looks for its survival value for the species (pp. 8–9). As an ethologist, he denies that all behaviour is conditioned by the environment (p. 41), and devotes himself to studying instinctive behaviour patterns. What is distinctive of these is that they do not always need an external stimulus to set them off, but happen spontaneously, as if driven by causes within the animal itself. Thus a male dove deprived of its mate would begin to perform its courtship dance to a stuffed pigeon, a piece of cloth, or even the empty corner of its cage (p. 42). And a hand-reared starling which had never caught flies or seen any other bird do so would go through fly-catching movements even when no flies were there (p. 43). Lorenz holds that there are many such patterns of animal behaviour which are 'hereditary co-ordinations' or 'instinct movements'; they are innate rather than learned, and for each there is a 'drive' which causes the behaviour to appear spontaneously (p. 74). But he also suggests, somewhat vaguely and tentatively, that such fixed action patterns are often at the disposal of one or more of the 'four big drives'–feeding, reproduction, flight, and aggression (p. 75). He thinks that any one piece of behaviour is usually caused by at least two drives or inner causes (pp. 73, 84), and that conflict between independent impulses can give firmness to the whole organism, like a balance of power within a political system (p. 80).

In *On Aggression* Lorenz devotes most of his attention to the natural history of aggressive behaviour, which he believes to be instinctive, driven by one of the major drives. He is concerned with fighting and threats between members of the *same* species, not with the attack of predator on prey, the mobbing of predators by prey, or the self-defence of any cornered animal (pp. 18–22). Concentrating thus on intra-specific aggression, he asks what its species-preserving function can be, and comes up with several answers. Firstly it can space out the individuals of a species evenly over the available territory, so that there is enough food for each (pp. 24–30). On a coral reef, each kind of fish has its own peculiar source of food, and each individual will defend its 'territory' against others of the same species although it easily tolerates fish of other species. The setup is like a series of villages, in each of which there is a living for only one butcher, one baker, and one candlestick-maker. Secondly and thirdly, aggression between rival males of a species ensures that the strongest individuals leave offspring, and are available for defence of family and herd (p. 31). Lastly, aggression can serve to establish and maintain a 'pecking order' or hierarchy in an animal community, which can be beneficial in that the oldest and most experienced animals can lead the group and pass on what they have learned (pp. 35–7).

But how can intra-specific aggression have such survival value without leading to injury and death, which obviously contradict survival? The remarkable fact is that despite aggression being so widespread among vertebrate animals, it is rare for an animal to be killed or seriously injured in the wild by members of its own species. Much aggressive behaviour takes the form of threats or pursuits rather than actual physical combat. Lorenz theorizes that evolution has produced a 'ritualization' of fighting, so that it can produce the above advantages without actually causing injury (pp. 93–8). Especially in heavily armoured animals, which must cooperate for breeding and perhaps for hunting, there is a need for a mechanism by which aggression can be inhibited (p. 110). So typically there is some kind of appeasement gesture or ritual submission by which a weaker animal can inhibit the aggression of a stronger. Beaten dogs, for instance, offer their vulnerable neck to the jaws of their opponent, and this seems to activate some specific inhibition mechanism, for it is as if the victor *cannot* then bring himself to administer the fatal bite (pp. 113–14) but just accepts that victory has been conceded.

Theory of Man

Lorenz sees man as an animal who has evolved from other animals. Just as our bodies and their physiology show a recognizable continuity with those of other animals, so Lorenz expects our behaviour patterns to be fundamentally similar to those of animals. To think of ourselves as different in kind, whether in virtue of free will or anything else, is an illusion. Our behaviour is subject to the same causal laws of nature as all animal behaviour (pp. 190–2, 204, 214), and it will be the worse for us unless we come to recognize this. Of course, we are different in *degree* from the rest of the animal world, we are the 'highest' achievement so far reached by evolution (p. 196). To explain our behaviour causally does not necessarily take away from our 'dignity' or 'value,' nor does it show us not to be free, for increasing knowledge of ourselves increases our power to control ourselves (pp. 196–202). Though Lorenz does not take discussion of these philosophical questions very far, he shows himself much more sensitive to them than Skinner.

The crucial point of Lorenz's view of human nature is the theory that like many other animals we have an innate drive to aggressive behaviour towards our own species. He thinks that this is the only possible explanation of the conflicts and wars throughout all human history, of the continuing unreasonable behaviour of supposedly reasonable beings (pp. 203–4). He suggests that Freud's theory of the death instinct is an interpretation of the same fundamental fact of human nature (p. 209). Lorenz seeks an evolutionary explanation for our innate aggressiveness, and for its peculiarly *communal* nature (for the most destructive fighting is not between individuals but between groups). He speculates that at a certain stage of the evolution of our ancestors, they had more or less mastered the dangers of their non-human environment, and the main threat came from other human groups. So the competition between neighbouring hostile tribes would have been the main factor in natural selection, and accordingly there would be a survival value in the 'warrior virtues' (p. 209). Selection can thus determine the evolution of cultures as well as of species (p. 224). At this postulated prehistoric stage, those groups that banded together best to fight other groups would tend to survive longest. Thus Lorenz explains the existence of what he calls 'militant enthusiasm,' in which a human crowd can become excitedly aggressive and lose all rationality and moral inhibitions (pp. 231–5): it has evolved from the communal defence response of our pre-human ancestors (p. 232).

Diagnosis

'All the great dangers threatening humanity with extinction are direct consequences of conceptual thought and verbal speech' (pp. 204–5). Thus our greatest gifts are very mixed blessings. Men are omnivorous creatures, physically quite weak with no great claws, beak, or teeth, so it is quite difficult for one man to kill another in unarmed combat. Accordingly there was no evolutionary need for strong inhibition mechanisms to stop fighting between ape-men. The heavily armed carnivores have such mechanisms (p. 207), but other animals do not; this explains why the dove–the very symbol of peace–can uninhibitedly peck to death a second dove which is enclosed in the same cage and cannot escape (see *King Solomon's Ring,* p. 184). But cultural and technological development puts artificial weapons in our hands–from the sticks and stones of pre-human ancestors, through the arrows and swords of history, to the bullets and bombs of today. The equilibrium between killing potential and inhibition is upset (p. 207). Thus Lorenz explains how it is that human beings are the only animals to indulge in mass slaughter of their own species.

Appeals to rationality and moral responsibility have been notoriously ineffective in controlling human conflict. Lorenz explains this by his theory that aggression is innate in us–like the instincts in the Freudian id, it must find an outlet in one way or another. Reason alone is powerless, it can only devise means to ends decided on in other ways, and it can only exert control over our behaviour when it is backed by some instinctual motivation (p. 213). So, like Freud, Lorenz sees a conflict between the instincts implanted in us by evolution, and the moral restraints necessary to civilized society. He speculates that in pre-human groups there must have been a primitive morality which condemned aggression within the tribe (pp. 215–16), for any tribe which fought within itself would soon lose the competition with other tribes. But the pressures of that competition produced an instinct for aggression against other tribes. Thus our technology of weapons has far outstripped the slow development of appropriate instinctive restraints on their use, and we find ourselves in the highly dangerous situation of today, with both the power to destroy the world and the *willingness* to do so in certain situations.

Prescription

If aggression really is innate in us, then there might seem to be little hope for the human race. For we have seen the uselessness of mere appeals to reason and morality, and if we try to eliminate all stimuli that provoke aggression, the inner drive will still seek outlets. Theoretically, we could try to breed it out by deliberate eugenic planning of human reproduction. But even if this were morally and politically possible, Lorenz thinks it would be highly inadvisable, since we do not know how essential the aggressive drive may be to the makeup of human personality as a whole (p. 239). If we eliminate aggression we might destroy at the same time many of the highest forms of human achievement.

Nevertheless, Lorenz avows optimism in his final chapter, and believes that 'reason can and will exert a selection-pressure in the right direction' (p. 258). For the more we begin to understand the natural causes of our aggression, the more we can take rational steps to redirect it. Self-knowledge is the first step to salvation (another echo of Freud, Sartre, and Socrates!). The next is sublimation, the redirection of aggression to substitute objects in harmless ways (p. 240). We can smash cheap crockery to express rage, and we can channel group-competitiveness into team games. We must break down mistrust between human groups by promoting personal acquaintance between individuals of different nations, classes, cultures, and parties (pp. 243–4). And we must redirect our enthusiasm to causes which can be genuinely universal among all peoples—Art, Science, and Medicine (pp. 244–9). Lastly, Lorenz avows great confidence in the human sense of humour, as something which promotes friendship, attacks fraud, and releases tension without getting out of rational control (pp. 253–7). So humour and knowledge are the two great hopes of civilization. In such means he sees hope that in future centuries our aggressive drive can be reduced to a tolerable level without disturbing its essential function (pp. 257–8).

Critical Discussion

Lorenz's theory is a persuasive one, for he seems to combine the insight of Freud with the scientific rigour of Skinner. Nevertheless there are important doubts to be raised about his theory and diagnosis of man. Unless we are professional researchers into animal behaviour, we can hardly be qualified to discuss the details of Lorenz's theories. Some of his factual claims about certain species—for instance, the alleged 'bloody mass-battles' of the rat (pp. xi, 139)—have been disputed. Here of course we need the facts to be ascertained. What we *can* begin to discuss without leaving our armchairs is the methodology of postulating instincts or inner drives to explain behaviour. We found this to be one of the weakest parts of Freud's theories, yet we could not agree with Skinner's total rejection of such postulation. Has Lorenz found the right middle path between these extremes? The crucial question is whether his application of the concepts of drive and instinct is falsifiable by observation and experiment. Now when he postulates a drive to explain a specific fixed-action pattern in a particular species–like the fly-catching routine of the starling–there do seem to be clear tests of the proposition. We can establish that a given action pattern is innate by showing that all normal individuals of the species of the relevant age and sex will perform the action, without previous learning from other individuals or from trial and error. If we also find that the stimulus which usually releases the action does not always do so with the same effectiveness (mating behaviour varies with the season) and if we also find that the action can sometimes be produced by less than the usual stimulus (like the isolated dove which courts the corner of its cage), then it is reasonable to say that there is some internal driving factor which varies in its strength.

So the presence of varying drives for specific fixed-action patterns is testable. But what is more dubious about Lorenz's methodology is his suggestion that such 'little partial drives' are often at the disposal of one or more of the 'four big drives' (feeding, reproduction, flight, and aggression; pp. 74–6). He holds that a 'self-contained function' is never the result of one single drive (p. 73), and even suggests that aggression is one of the driving powers which 'lie behind behaviour patterns that outwardly have nothing to do with aggression, and even appear to be its very opposite' (p. 35). On the face of it, this would seem to permit us to attribute any kind of behaviour at all to aggression, and thus make such attribution untestable and unscientific. (It is exactly parallel to Freud's theory of 'reaction formation,' by which an inner tendency can be expressed in the opposite behaviour.) There *may* be ways of testing such talk of basic drives, intermingling of drives, and diversion of drives to different behaviour, but until the testability is shown, such theorizing is not scientific. And until the tests give confirmation, there is no reason to suppose it true.

Apart from these methodological questions about the general theory, there must be considerable doubt about the way in which Lorenz argues from animals to men. (This was also a major defect of Skinner.) In *On Aggression* Lorenz takes most of his examples from fish and birds, few from mammals, and hardly any from our closest relatives, the great apes. Yet he is prepared to argue by analogy that if fish and birds are innately aggressive, then human behaviour is subject to the same basic laws (p. 204). The analogy must surely be judged a weak one. It would be stronger if he had made detailed studies of chimpanzees and gorillas, as recent researchers like Jane van Lawick-Goodall have done (see her *In the Shadow of Man*). But even evidence about these nearer relations is very far from showing us the essential nature of man, although popular writers such as Morris would have us believe so. For the *differences* between men and other animals may be as important as the similarities. In general, to show that X has evolved out of Y does not show that X *is* Y, or is nothing but Y, or is essentially Y. Even if it could be shown that sectarian conflict (e.g., in Ulster) is evolved from the territorial defence mechanisms of tribes of ape-men, this still does not show the former to be nothing but the latter. In any case, theories of pre-human behaviour, such as Lorenz's suggestions of what competition between hostile tribes must have been like, are highly speculative, and it is hard to see how we could now find any evidence for or against them.

These doubts must therefore infect the crucial feature of Lorenz's theory of human nature—the idea of innate aggression. For if the analogy from animals does not prove this, we must make direct observation of human behaviour to test it. At this level, Lorenz is as amateur as the rest of us who are not anthropologists or sociologists. We must look not to his speculations but to the facts. Anthropologists have described some societies in which aggression is notably absent (see some of the papers in *Man and Aggression*—mentioned below). This would suggest that it is more socially learned than innate. In modern industrial societies it does seem that overt violence varies somewhat according to social background. No doubt it will be suggested that middle-class economic competition is just as 'aggressive' as working-class gang warfare; but then the term is being extended to cover more than physical violence and the threat of it. Clearer definition of the term is a prerequisite of further inquiry; and that inquiry looks like being at least as much sociological as biological. We must judge Lorenz's theory of man as a speculative generalization from his observation of animals. But it points us to a vitally important area for research into human nature.

For Further Reading

Basic text: *On Aggression,* translated by Marjorie Latzke (Methuen, London, University Paperback 1966); translated by Marjorie K. Wilson (Bantam Books paperback, New York, 1974).

The two volumes of Lorenz's *Studies in Animal and Human Behaviour* (Methuen, London, 1970 and 1971; Harvard University Press, Cambridge, Mass., 1970 and 1971) give more details both of his ethological studies and of the philosophy of science underlying them. Although more technical than the above, these are still intelligible without previous knowledge of science or ethology.

For criticism of Lorenz and other ethological writers, see *Man and Aggression,* edited by M. F. Ashley Montagu (Oxford University Press Galaxy Books paperback, New York, 2nd edn. 1973).

For Darwin's theory of evolution, see his *Origin of Species,* reprinted in Pelican Classics, 1968, and in a Mentor paperback (New American Library, New York).

Since this chapter was written, 'sociobiology' has burst upon the scene. See E. O. Wilson, *Sociobiology: The New Synthesis* (Harvard University Press, Cambridge, Mass., 1975), especially the last chapter on man, and his *On Human Nature* (Harvard University Press, Cambridge, Mass., 1978). For a philosophical overview of this controversial territory, see M. Ruse, *Sociobiology: Sense or Nonsense?* (Reidel, Dordrecht, 1979). See also notes 3 and 4 to Chapter 10.

Page references in this chapter are to the Methuen edition of *On Aggression*. Readers of the Bantam Books edition should use the references below.

Stevenson	*Bantam Books paperback*
page 123 refers to pages	9–11; 47–8; 48–9; 49–50; 83–4
124	85; 82; 94; 90; 21–6; 28–35; 35–6; 41–3
125	104–10; 123; 127; 214–16, 229, 240; 220–1; 221–7; 228–9
126	235; 235; 251; 259–64; 261; 230; 233
127	233; 240; 241–2
128	268–9; 290; 269–70; 273–4; 274–80; 283–8; 289–90; xi–xii, 157
129	83–6; 82; 40
130	229

III

Conclusion

10

Some Lines for Further Inquiry

If the reader is expecting this book to conclude with some all-embracing truth about human nature, then he or she will be disappointed. I have no eighth Stevensonian 'theory' to offer, merely an invitation to further inquiry in a variety of directions.

Although we have treated our seven theories as if they were rivals for our allegiance, they will hardly be incompatible with each other on all points. Unless one has a very exclusive kind of commitment to a framework of thought—the sort of total and exclusive allegiance to a 'closed system' we discussed in Chapter 2—one can see each theory as emphasizing (though perhaps *over*-emphasizing) different aspects of the total truth about human nature. Each of these views has surely made *some* positive contribution to our understanding of ourselves and our place in the universe. For example, there is no difficulty in principle in allowing for the influence of *both* Marxist socio-economic factors and biologically innate factors, perhaps of a Freudian kind, in the formation of culture and individual character. In this way, one might begin to see the rival theories as adding up, rather than cancelling out.

But of course we must recognize that there are some important differences between them, both in their specific claims about human nature, and in their various background metaphysical and methodological assumptions (which I summarized briefly under the catchall label 'theory of the universe'). In this final chapter, I will point to some of the main issues that arise from our discussion, and make some suggestions for where to look for further enlightenment on them (I have selected a number of books, which are mentioned in the notes at the end of this chapter).

One enormously important truth about human beings is that we have evolved from simpler forms of life by natural selection over vast tracts of time. Early in Chapter 9 I said that the huge mass of evidence makes this evolutionary origin of man indisputable, 'an established fact'; yet one cannot say that it is not disputed, for 'creationists' in the United States have recently reopened nineteenth-century debates about this. The very fact that it is thus contested is interesting, however, for it shows just how difficult it is to maintain scientific objectivity when discussing human nature. As we have noted, those with firmly rooted religious or political beliefs will not be ready to change them just because some scientific theory is said to go against them; they will typically challenge the scientific evidence and its interpretation. But of course I cannot rationally answer the creationists' assertions just by diagnosing ideological motives—that would be to put myself in danger of maintaining Darwinian theory as a 'closed system' in just the way condemned in Chapter 2. What needs to be done is to consider each of the objections to evolutionary theory in detail, and to show that in the light of all available evidence it does not stand up. This is hardly the place to begin doing that, for the job has been very effectively done by others.[1]

I would not want to suggest that evolutionary theory is without problems: no scientific theory can enjoy complete certainty, for history shows us how all such theories develop and change. But as I understand it, there is no serious present competitor to a basically Darwinian story about man. Any adequate understanding of human nature must therefore take into account this evolutionary origin. What I have in mind here is not just that we must admit this (as the majority of Christian theologians have done for a century) but that we must reckon with the possibility that our evolution may *explain* much about human nature. Several recent books discuss the implications of this idea. E. O. Wilson, the recent founder of 'sociobiology,' sets out the general case for such an approach and applies it to some specific categories of human behaviour—aggression, sex, altruism, and religion—in a book simply entitled *On Human Nature.*[2] Other biologists have hotly criticized both some of the scientific claims of sociobiology and what they see as the reactionary social and political pronouncements that have been made on this supposedly scientific basis.[3] Philosophers have also begun to consider the evolutionary approach, criticizing its excesses, but also admitting its potential explanatory power.[4]

The general idea is that, like any other animal species on this planet, we have a certain genetic constitution that causally explains not only the anatomical features which distinguish our bodies from those of other species, but also our distinctive human *behaviour,* especially our social, other-directed behaviour.

That there are *some* such innate tendencies is indisputable—for example, sexual behaviour is rooted in our biological nature. But even that obvious example immediately raises problems and questions, for the forms sexuality takes vary considerably between societies, and in devotedly celibate individuals like monks and nuns its expression may be suppressed. We have some innate biological drives, certainly, but we seem to be unique in the way in which our behaviour depends on the particular human culture we are brought up in (rudimentary cultural differences have been discerned in some of the higher animal species, but to nothing like the human extent). Thus theorists like Lorenz and Wilson who inquire into human ethology or sociobiology have been met with storms of protest from those who maintain that, apart from the most obvious biological universals like eating, sleeping, and sex, human behaviour depends on culture rather than biology. The objectors tend to suspect ideological motives behind assertions that such-and-such behaviour, for example, aggression or competition, is innate in human biological nature, for they see a danger of such claims being used to justify as 'natural' certain social practices, for example, war and preparation for war, or capitalist economic systems, which, they would want to argue, need not be inevitable at all if certain social features were changed. But equally, of course, there may well be political motives behind some of the *resistance* to sociobiological claims. In this case, too, what is needed is the hard work of investigating in detail the evidence for those claims.

The general contrast between the innate and the learned, heredity and environment, nature and nurture, has come up in this book at various points. Plato, Marx, and Skinner emphasize, as we have seen, the power of social conditioning and the possibilities of changing individuals by reforming social structures and practices. Sartre, in his highly individualistic early theory, claims we each have the power to change ourselves radically. Freud and Lorenz emphasize the limits to such change in the innate universal nature of humankind. Christian doctrine warns us that no real change, individually or socially–no escape from the fundamental problem of sin–is possible without the grace of God.

To see how the issues could be pursued in more detail, let us review briefly those four topics which Wilson discusses. Firstly, sexuality poses social problems as well as questions for philosophical exploration and scientific explanation. For instance, homosexuality arouses strong feelings and social controversy, but it is also rather puzzling biologically–how could tendencies to non-reproductive sex get reproduced? The phenomenon challenges our assumptions about what is 'natural.' A similar challenge is posed by the feminist movement, which has in recent years raised important questions about how far the differences between men and women are really a matter of innate, biological, 'natural' tendencies, or how far they are produced by the cultural conditioning of various societies, mostly serving the interests of men. One point of entry to the burgeoning literature on this set of issues is the chapter on women's nature and human nature in Jean Grimshaw's recent book; more detailed discussion is to be found in Alison Jaggar's systematic treatise.[5]

Secondly, aggression is something which manifestly poses problems for us–for the very survival of the human species on earth is now threatened by the number of nuclear weapons ready to explode at short notice. But it is not an easy phenomenon to understand and explain. Indeed, as we noted at the end of Chapter 9, the term itself is too vague (Wilson distinguishes seven different categories of aggression in animals). The Freud-ian/Lorenzian hypothesis of a specific aggressive drive or in-stinct constantly seeking release now looks oversimplified, at best. A more plausible view, which allows a crucial role for social environment, is that we are predisposed by our genes to become highly aggressive towards each other in certain so-cial conditions.

Defining what these social situations are must surely involve the third topic, which sociobiologists label 'altruism.' They thereby divert the word somewhat from its ordinary meaning and apply it to any animal behaviour that appears to favour another individual's chances of survival and reproduction. On the face of it, it is hard to explain in strictly Darwinian terms any human behaviour which is not directed to the person's own survival, or that of his or her children, or other near relations who carry some of his or her own genes. Yet human beings manifestly do things, involving various degrees of sacrifice, for all sorts of wider social groups–schools, corporations, sports teams, tribes, racial, ethnic, and religious groups, nations–up to and including the 'supreme sacrifice' of one's life for one's country in wartime. Indeed (as Lorenz noted) the most dangerous kind of human aggression is precisely when specific-ally *communal* hostility is involved. Down through history tribes, races, and nations have in a collective way done the most terrible things to each other; and every day the news bul-letins give more instances from all around the world.

But the new fact at the level of nations is that science has increased our military destructive power to the point where, although we are frightened to use it, we are constantly in peril of extermination.[6] There may be hope (as Wilson sug-gests) in the fact that the groups between which hostility exists can change quite quickly–for example, the present superpower confrontation has gone on for only forty years, which is not long in historical terms. But there is immense danger in the fact that from now on, whenever the dividing line between 'us' and 'them' happens to be, technical knowledge offers each side the power to destroy not just all of 'them' but very likely all of 'us' too, and the rest of the living world besides. The disturbing studies of Milgram[7] confirm by experiment what is already well known from experience—that in obedience to figures they see as 'authoritative,' people will do to 'others' what they would otherwise do to nobody. Glenn Gray's deep reflection on his wartime experience shows how war can bring out both the best and the worst in human nature.[8] Given the extremity of our peril, nothing can be more urgent than better understanding of our tendency to intercommunal aggression.

There is a connection with religion, the fourth topic on our list, in that typically the hostility between groups is expressed with ideological intensity, and very often it takes an explicitly religious form, with the other side being categorized as unbelievers, infidels, infected with evil. Clearly, religion has immense emotional power over humankind, compared with the purely rational approach of science and philosophy. Even for those who (with Freud and Marx) would say that the essential content of all religious beliefs is illusory—postulating some sort of personal power beyond this world who cares in some way about our lives—there is an obvious question, namely, why is it that all known human societies suffer from some form of this illusion? Can it be explained in terms of psychoanalytic theory (the wish for a father-figure), or in sociological terms as Marx and Durkheim claimed, or in evolutionary (ultimately genetic) terms, as Wilson suggests? These various reductive explanations are intellectually interesting, but are any of them really convincing? Is not religion an attempt to cope with problems which confront *any* human being, however comfortably situated, in however reformed a society, how to face up to the inevitability of one's own death, how to cope with the gap between one's potential and one's achievement: why go on living? what is the purpose? Yet the philosophical question of whether the essential content of religious assertions *is* illusory must be faced. We began to raise this question in the chapter on Christianity but did not solve it, of course—all I can do now is to recommend the reading suggested there.[9] Assessing the truth-value of religious claims is one route into central philosophical inquiries about meaning, epistemology, and metaphysics. Religion is thus one important area where psychological or sociological explanations depend on philosophical assumptions about the meaning and truth of the beliefs to be explained.

One general lesson we have learned from this book is that when we theorize about human nature the application of scientific method to ourselves is not as straightforward as enthusiasts like Skinner would have us believe. For one thing, there is the matter of ideology (touched on briefly in Chapters 1 and 2 and again here, and raised in a more theoretical way in the chapter on Marx). Every human way of life presupposes some beliefs about human nature, and when a belief is thus involved in their thought and action, people will typically resist changing it. But proponents of change will often appeal to their own claims about human nature. So whenever social scientific evidence is adduced for such claims–for example, about mental differences between racial groups, about the docility of women or the aggressivity of men–we must be alert to the possibility that such assertions serve the interest of certain groups rather than others, and that the factual evidence, such as it is, only shows how the relevant people have so far tended to behave in certain forms of society, which we may now want to change. In examining various theories here, we have tried to show what *values* are involved in their diagnosis and prescriptions for humankind, and sometimes in their 'basic' theory of human nature.

The other main sort of difficulty in theorizing scientifically about human nature is the question of how far human actions and thoughts, because of the mental aspects involved, are amenable to scientific explanation at all. A host of difficult philosophical problems arise here. Someone who believes, like Plato and Descartes, that we are essentially non-material souls is going to see our most distinctively human nature as beyond all scientific investigation. This metaphysical issue of dualism or monism has to be faced. Is man made of matter alone, or is consciousness necessarily non-material in nature? In what way, if any, is it possible for a person to survive death? Are mental states (sensations, emotions, beliefs, desires, etc.) and brain states (the electrical and chemical goings-on investigated by neurophysiologists) two different sorts of thing, or just two aspects of one set of events?[10] Of the theories considered in this book, only Plato's is unambiguously dualist; we found that Christianity does not seem to need the notion of disembodied soul.

But even non-dualists tend to think there is something distinctive about human action and thought–its freedom and rationality–which makes straightforward scientific explanation in terms of causes and laws of nature somehow problematic or inappropriate. The traditional way of raising this issue is whether and how there is room for free will in a world of determining causes, but more recently this has been seen as part of another question–whether and how there is room for *rationality,* for people to have reasons for their beliefs and actions. This set of issues came to the surface at the end of our chapters on Freud and Skinner, and is presented in a particularly dramatic, but idiosyncratic, way by Sartre. They are at the centre of contemporary philosophy of mind and action, and are fundamental for psychology and all study of human nature, since they ask what sort of explanations of human mental phenomena it is possible to achieve.[11]

Lastly, in a book which presents *theories* it is well worth reminding ourselves how much understanding of human nature we have, and can gain, in a non-theoretical way from our experience of particular cases. Foremost of course comes our own experience of life, of the individual people we have had dealings with, and of the communities and cultures we have lived in. But the study of history and ethnography can extend our acquaintance beyond our own limited experience, to individuals and societies with their own particular characters, but distant in time or space.[12]

Let us also remember how literature presents us with imaginary, but in another sense very 'real,' particular cases of men and women displaying their human nature in feeling, thought, and action. In the great works, our understanding of human nature is extended and deepened, though not necessarily in ways we can put into explicit words. To pick an obvious, but unmissable, example—consider the plays of Shakespeare, with the aid of perceptive commentators.[13] Correspondingly, alas, other works of fiction which enjoy much wider, although shorter-lived, popularity—whether in the media of print, film, or television—can vividly present shallow, stereotyped characters and social situations, and thus restrict rather than extend our understanding of human nature. My point, however, is that understanding *can* be deepened by attention to particular cases, whether encountered in one's own experience, reported from the real world, or presented imaginatively in the great works of literature.

Human nature is a topic which breaks down the boundaries between the sciences and what have been called 'the humanities.' Our urgent social and political problems worldwide cry out for better understanding of human nature. How often is it the case that the technical problems involved are soluble but what seem insuperable are the political and social obstacles? More than ever before, the proper study of mankind is man.

Notes

1. See P. Kitcher, *Abusing Science: The Case against Creationism* (MIT Press, Cambridge, Mass., 1982; Open University Press, 1983).
2. E. O. Wilson, *On Human Nature* (Harvard University Press, Cambridge, Mass., 1978).
3. Steven Rose, R. C. Lewontin and Leon J. Kamin, *Not in Our Genes: Biology, Ideology and Human Nature* (Penguin, London, 1984).
4. Mary Midgley's *Beast and Man* (Methuen, London, 1980) discusses the general approach—and Wilson's specific topics—in the light of moral philosophy. F. von Schilcher and N. Tennant (biologist and philosopher, respectively) in *Philosophy, Evolution and Human Nature* (Routledge & Kegan Paul, London, 1984) concentrate more on biology, epistemology, and language.
5. Jean Grimshaw, *Feminist Philosophers: Women's Perspectives on Philosophical Traditions* (University of Minnesota Press, Minneapolis, 1986; Wheatsheaf Books, Brighton, 1986); Alison

Jaggar, *Feminist Politics and Human Nature* (Rowman and Allanheld, Totowa, N.J., 1983; Harvester Press, Brighton, 1983).

6. See *The Fate of the Earth* by Jonathan Schell (Knopf, New York, 1982; Picador, London, 1982) for an eloquent meditation on the enormity of this new phenomenon for humankind.
7. Stanley Milgram, *Obedience to Authority* (Harper & Row, New York, 1975; Tavistock, London, 1974).
8. J. Glenn Gray, *The Warriors: Reflections on Men in Battle* (first published in 1959; Harper & Row, New York, 1970).
9. See pp. 51–2; and for some eloquent sermons expounding biblical passages in the light of the sorts of 'existential' questions just mentioned, see Paul Tillich, *The Shaking of the Foundations* (first published in 1949; Penguin, London, 1962).
10. Jenny Teichman's little book, *The Mind and the Soul* (Routledge & Kegan Paul, London, 1974), is one useful introduction to this and other main issues in the philosophy of mind.
11. Kathleen Wilkes' book *Physicalism* (Routledge & Kegan Paul, London, 1978) is one of the clearest of many that discusses these questions further, and D. C. Dennett's *Elbow Room: The Varieties of Freewill Worth Wanting* (Oxford University Press, New York, 1984) is a lively recent treatment of the free will problem.
12. One book that comes to mind here is *The Identity of Man* (Methuen, London, 1983), by the archeologist Grahame Clark, which emphasizes the amazing diversity of past cultures on this planet. *The Perfectibility of Man,* by John Passmore (Duckworth, London, 1970) is a fascinating tour through the history of Western ideas about human nature.
13. See, for example, John Wain, *The Living World of Shakespeare* (Macmillan, London, 1964); Germaine Greer, *Shakespeare* (Oxford University Press, New York, Past Masters Series 1986).

Index